# Toward the Future: The New Challenges of the Cell Therapy and Potential of Regenerative Medicine

## Edited by

### Nicola Daniele

*CryoLab, Centre of Biotechnology and Cryobiology*
*Rome, Italy*

### &

### Francesco Zinno

*Faculty of Medicine and Cryolab University*
*of Rome "Tor Vergata" Rome, Italy*

*And*

*Azienda Ospedaliera di Cosenza*

advertisements or ideas contained in the Work.

## *Limitation of Liability:*

In no event will Bentham Science Publishers, its staff, editors and/or authors, be liable for any damages, including, without limitation, special, incidental and/or consequential damages and/or damages for lost data and/or profits arising out of (whether directly or indirectly) the use or inability to use the Work. The entire liability of Bentham Science Publishers shall be limited to the amount actually paid by you for the Work.

## General:

1. Any dispute or claim arising out of or in connection with this License Agreement or the Work (including non-contractual disputes or claims) will be governed by and construed in accordance with the laws of the U.A.E. as applied in the Emirate of Dubai. Each party agrees that the courts of the Emirate of Dubai shall have exclusive jurisdiction to settle any dispute or claim arising out of or in connection with this License Agreement or the Work (including non-contractual disputes or claims).
2. Your rights under this License Agreement will automatically terminate without notice and without the need for a court order if at any point you breach any terms of this License Agreement. In no event will any delay or failure by Bentham Science Publishers in enforcing your compliance with this License Agreement constitute a waiver of any of its rights.
3. You acknowledge that you have read this License Agreement, and agree to be bound by its terms and conditions. To the extent that any other terms and conditions presented on any website of Bentham Science Publishers conflict with, or are inconsistent with, the terms and conditions set out in this License Agreement, you acknowledge that the terms and conditions set out in this License Agreement shall prevail.

**Bentham Science Publishers Ltd.**
Executive Suite Y - 2
PO Box 7917, Saif Zone
Sharjah, U.A.E.
Email: subscriptions@benthamscience.org

**BENTHAM SCIENCE**

# CONTENTS

# FOREWORD

As a woman first, and as passionate researcher second, I was always attracted by what did represent Marie Curie and her breakthroughs for the entire scientific world. One of my favorite quotes is:

*"We must not forget that when radium was discovered no one knew that it would prove useful in hospitals. The work was one of pure science. And this is a proof that scientific work must not be considered from the point of view of the direct usefulness of it. It must be done for itself, for the beauty of science, and then there is always the chance that a scientific discovery may become like the radium a benefit for mankind".*

**Marie Curie**

For me this means that as researcher we must find passion in what we make to trigger the "out of the box" thinking and to realize the unimaginable. When I was a student I meet two distinguished researchers *Francesco Zinno*, and *Nicola Daniele* whom introduced me with the main goals of the hematopoietic stem cells (HSCs) manipulation processes. Cell therapy helped the mass implementation of one the breakthrough that changed the way of making medicine, the regenerative medicine. Regenerative medicine involves the use of some of the most advanced therapeutic technologies of the 21st century. We can define this branch of medicine as methods to replace or regenerate human cells, tissues or organs in order to restore or establish normal function with the use of cell therapies, tissue engineering, gene therapy and biomedical engineering techniques [1]. The rapid pace of the supporting science is likely to see its application across ever increasing fields of clinical practice. This book's main aim is to ponder on the importance of the regenerative medicine under the success that this area was given by stem cell research, and cell- and gene-based therapy. To start this journey, authors decided to give a general picture of the biology of human stem cells and their classification, to introduce the reader with their unique and extraordinary properties. After that, they continued with a typology-based structured analysis of the potentiality of these cells to upgrade the therapeutic treatments.

They explain better the actual use of the stems cells and stems cells-like for regenerative therapeutic purpose such as:

• Cell-therapy approaches in neurologic areas such as Parkinson's disease;
• Retinal repair;
• Optic nerve regeneration;
• Reconstruction of the tissues in cases of burn;

- Bladder repair;
- Liver diseases;
- Angiogenesis and cardiac repair.

Authors continued with the researches that aim to reach and overstep the actual bounders of the "personalized medicine" to underline the stage of evolution and the infinite possibility of this new outstanding way of thinking on medicine. This book outline an attractive promotion of the present and future benefits of mankind from the discovery of stems cells and their use on the regenerative medicine. The more we know about stems cells the more we approach with the potentiality of human body to heal its self.

**Entela Shkëmbi**
Semiology Laboratory Manager
Dani Infertility Clinic
Tirana
Albania

## REFERENCES

[1] "Building on our own potential: a UK pathway for regenerative medicine" A report from the Regenerative Medicine Expert Group.

# PREFACE

This eBook entitled "Toward the Future: The New Challenges of the Cell Therapy and Potential of Regenerative Medicine" comprises chapters written by the leading experts in this field that provide state-of-the art information about the developments in important selected areas of Regenerative Medicine.

Regenerative Medicine is the process of creating living, functional tissues to repair or replace tissue or organ function lost due to age, disease, damage, or congenital defects.

This field holds the promise of regenerating damaged tissues and organs in the body by stimulating previously irreparable organs to heal themselves.

Regenerative Medicine also empowers scientists to grow tissues and organs in the laboratory and safely implant them when the body cannot heal itself.

Importantly, Regenerative Medicine has the potential to solve the problem of the shortage of organs available through donation compared to the number of patients that require life-saving organ transplantation.

Regenerative Medicine is also one of the fastest growing biomedical industries in the world because patients are being cured of diseases that were once incurable. Moreover, this field represents a new paradigm in human health because the vast majority of treatments for chronic and life-threatening disease focus on treating the symptoms, not curing the disease.

In fact, there are few therapies in use today capable of curing or significantly changing the course of a disease.

Stem cell therapy, when combined with immune and gene therapy, shows even greater potential to cure diseases. This new combination of regenerative cell therapies will open a new age of medicine, forever changing how it is practiced.

We would like to express our gratitude to all the Authors for their excellent contributions. We would also like to thank the entire team of Bentham Science Publishers, particularly Dr. Humaira Hashmi and Prof. Atta-ur-Rahman for their excellent efforts. We are confident that this Volume will receive wide appreciation from students and researchers.

***Nicola Daniele***
*Editor*
Cryolab

University of Rome "Tor Vergata"
Rome
Italy

*Francesco Zinno*
*Editor*
Faculty of Medicine and Cryolab
University of Rome "Tor Vergata"
Rome
Italy
AND
Azienda Ospedaliera di Cosenza

# List of Contributors

**Alexander Aleynik**    Graduate School of Biomedical Health Sciences, Rutgers Univ., Newark, NJ, USA

**Carla Cerri**    Pediatric Oncology Hematology Unit, Perugia General Hospital, Località Sant'Andrea delle Fratte, 06156 Perugia, Italy

**Elena Mastrodicasa**    Pediatric Oncology Hematology Unit, Perugia General Hospital, Località Sant'Andrea delle Fratte, 06156 Perugia, Italy

**Federica Sangiuolo**    Department of Biomedicine and Prevention, Tor Vergata University, 00133 Rome, Italy

**Federica Tomassetti**    CryoLab, Centre of Biotechnology and Cryobiology, Rome, Italy

**Filippo Zambelli**    Vrije Universiteit Brussel, Research Group Reproduction and Genetics, Brussels, Belgium and S.I.S.Me.R. Reproductive medicine Unit, Bologna, Italy

**Francesco Arcioni**    Pediatric Oncology Hematology Unit, Perugia General Hospital, Località Sant'Andrea delle Fratte, 06156 Perugia, Italy

**Francesco Zinno**    CryoLab, Centre of Biotechnology and Cryobiology, Rome, Italy and Immunohaematology Section, Tor Vergata University, Rome, Italy, and Az. Ospedaliera di Cosenza

**Fulvia Fraticelli**    CryoLab, Centre of Biotechnology and Cryobiology, Rome, Italy

**Giuseppe Novelli**    Department of Biomedicine and Prevention, Tor Vergata University, 00133 Rome, Italy

**Grazia Maria Immacolata Gurdo**    Pediatric Oncology Hematology Unit, Perugia General Hospital, Località Sant'Andrea delle Fratte, 06156 Perugia, Italy

**Ilaria Capolsini**    Pediatric Oncology Hematology Unit, Perugia General Hospital, Località Sant'Andrea delle Fratte, 06156 Perugia, Italy

**Jimmy Patel**    New Jersey Medical School, Dept of Medicine, Hematology/Oncology, New Jersey Medical School, USA

**Katia Perruccio**    Pediatric Oncology Hematology Unit, Perugia General Hospital, Località Sant'Andrea delle Fratte, 06156 Perugia, Italy

**Lucia De Santis**    San Raffaele Scientific Institute, Vita-Salute University, Dept Ob/Gyn, IVF Unit, Milan, Italy

**Maurizio Caniglia**    Pediatric Oncology Hematology Unit, Perugia General Hospital, Località Sant'Andrea delle Fratte, 06156 Perugia, Italy

**Nicola Daniele**    CryoLab, Centre of Biotechnology and Cryobiology, Rome, Italy

| | |
|---|---|
| **Paola Spitalieri** | Department of Biomedicine and Prevention, Tor Vergata University, 00133 Rome, Italy |
| **Pranela Rameshwar** | New Jersey Medical School, Dept of Medicine, Hematology/Oncology, New Jersey Medical School, USA |
| **Rita Vassena** | Clinica EUGIN, Barcelona, Spain |
| **Silvia Franceschilli** | CryoLab, Centre of Biotechnology and Cryobiology, Rome, Italy |
| **Valentina Rosa Talarico** | Department of Biomedicine and Prevention, Tor Vergata University, 00133 Rome, Italy |

# Toward the Future: The New Challenges of the Cell Therapy and Potential of Regenerative Medicine

2

**Toward the Future: the New Challenges of the Cell Therapy and Potential of Regenerative Medicine**

Editor: Nicola Daniele and Francesco Zinno

eISBN (Online): 978-1-68108-437-4

ISBN (Print): 978-1-68108-438-1

First published in 2017.

# CHAPTER 1

# Biology of Human Stem Cells

## Silvia Franceschilli[*]

*Cryolab, University of Rome "Tor Vergata", Rome, Italy*

**Abstract:** Stem cells are always regarded as cells with unique and extraordinary properties. SCs are able to self-renew and differentiate into specialized cells and this is a great advantage to maintain homeostasis in the body. The cells can divide with two different strategies and they are influenced by intrinsic and extrinsic factors of their microenvironment in which there are: the niche. The internal signals are represented by the genetic information of the cells, the external signals come instead from the microenvironment and they are physical or chemical signals. Stem cells are classified into embryonic stem cells and adult stem cells. Embryonic stem cells are derived from the inner cell mass of the blastocyst, these have great potential and over the years researchers have studied their properties and the importance of keeping them in appropriate culture conditions. Adult stem cells are found in a large number of tissues and have the very important role to replace damaged cells in living tissue. SCs can also be classified according to their potential, so they can be defined as totipotent, pluripotent, multipotent and they can differentiate respectively in a decreasing number of specialized cells of the body.

**Keywords:** Adult stem cells, Division, Embryonic stem cells, Mesenchymal stem cells, Multipotent stem cells, Niche, Pluripotent stem cells, Stem cells, Totipotent stem cells.

## INTRODUCTION

Stem cells (SCs) have unique features because they are an undifferentiated kind of cells, that have capacity to renew themselves and that can turn themselves into many different cells types with specific functions. SCs have the important role to

---

[*] **Corresponding author Silvia Franceschilli:** Cryolab, University of Rome "Tor Vergata", Rome, Italy; Tel/Fax: +0393384432929; E-mail: silvia.franceschilli@gmail.com

maintain the homeostasis in the body because in some organs they can renew, maintain or replace damaged tissues. In other organs they also can divide under special conditions [1]. The potential of SCs consists of dividing themselves without limits to renew cells and tissues, and the cells that resulted from this division can remain stem cells or become specialized type of cells [2]. The ability of SCs to "self-renew" or to generate differentiated cells can be defined by some signals derived from the special microenvironment of stem cells that is defined as "niche". In 1978, Schofield developed the hypothesis of this environment that could maintain the proprieties of stem cells [3, 4]. In the niche, cells can be influenced by internal and external signals to divide themselves by two different mechanisms: symmetric or asymmetric strategy [5]. Asymmetric division is characterized by the generation of a daughter with stem-cell fate and a cell that differentiates into different types. This mechanism is useful because of the production of two products with a single division but it can be considered a problem because it is not able to expand the number of stem cells. Over the years a lot of studies were carried out to describe the mechanisms that rule this type of division [6 - 9] and today it is possible to describe two different types of them: intrinsic or extrinsic mechanism. The second type of division is the symmetric one, that is defined by the ability of stem cells to divide themselves symmetrically to generate two stem cells or two differentiated cells. This type of division can be considered useful during wound healing and regeneration [10].

During the differentiation SCs lose their special condition of unspecialized cells, and this process is controlled by several steps and it is lead by internal and external signals. The internal signal consists of the information that is contained in the DNA and the external is represented by physical or chemical signals coming from the microenvironment. The combination of these elements regulates the behavior of stem cells [1].

There are different types of SCs, that can be classified into two groups: embryonic and adult stem cells [2]. According to their potential of differentiation, stem cells can be classified as totipotent, pluripotent or multipotent stem cells. Totipotent stem cells are capable to generate all the body because of their high capacity for differentiation. Pluripotent stem cells have the ability to form about 200 kinds of differentiated cells but not an organism, and finally multipotent stem cells can define cells of specific tissue. Haemopoietic stem cells can be considered multipotent stem cells because they can form any blood cells but not other tissues [11].

## Embryonic Stem Cells

In 1981 researchers discovered how to isolate embryonic stem cells from mouse embryos. At the end of nineties, in 1998, scientists described a method to derive stem cells from human embryos and how to grow them in culture [12 - 14]. They were defined human embryonic stem cells. In the early steps of embryogenesis, 3 or 5 days after the fertilization, the blastocyst is formed. It is composed by three parts, the trophoblast, an hollow cavity inside and finally the inner cell mass (ICM). This cell mass is composed by a special type of cells, that have the extraordinary potential to differentiate into a significant number of specialized cell types [15]. Embryonic stem cells can be maintained in culture in their undifferentiated state for a long time but they require appropriate conditions because gene expression and property of cells can be influenced by the environment [16, 17]. These cells are characterized by the expression of some specific transcription factors such as OCT3/4, NANOG and SOX2 [18]. The transcription factors OCT4, SOX2, NANOG control the expression of genes including other transcription factors such as STAT3, HESX1, FGF-2 and TCF. Moreover these transcription factors control signaling elements that are necessary to maintain the stem cell state and they repress some genes that would stimulate differentiation [19]. Embryonic stem cells retain the ability to differentiate themselves into many types of cells representing the three germ layers consisting of endoderm, mesoderm and ectoderm [20].

## Adult Stem Cells

Human tissues are able to adapt themselves to different environmental conditions and to use a certain plasticity to survive in different circumstances. This is possible thanks to the presence of adult stem cells [21]. These cells are undifferentiated and they are among differentiated cells in tissues or organs and they live in specific areas defined as "niches" [1 - 22]. Niche is sensitive to the different hormonal signals or those coming from the microenvironment and can direct the cells to a division and differentiation to ensure homeostasis. Adult stem cells are in a state of quiescence until it comes a stimulus that signals them to differentiate [21]. These cells can divide and differentiate to maintain the homeostasis of living parts of the body because they can replenish dying cells or damaged tissues [23]. Studies about adult stem cells begun in 1950s when researchers discovered special cells in the bone marrow. These cells have been identified in many tissues and organs such as bone marrow, brain, skin, heart, skeletal muscle, teeth, gut, liver, ovaries and testis [1 - 21]. Among adult stem cells, it is possible to distinguish their different types: hematopoietic stem cells characterized by specific surface markers, and stromal stem cells or mesenchymal stem cells. Hemopoietic stem cell are multipotent stem cell that are able to divide themselves to form blood cells. These cells have a very high rate of differentiation because every day billions of new blood cells are formed. Hemopoietic stem cells are characterized by the expression of some surface markers such as CD34, CD38, CD59, CD133 [24]. Mesenchymal stem cells are characterized by the expression of different surface markers but they do not express those that are specific of hematopoietic stem cells [25]. According to the criteria established by the International Society for Cellular Therapy, mesenchymal stem cells are identified by their ability to adhere to plastic when they are in culture, and they express surface markers such as CD29, CD44, CD90, CD49a-f, CD51, CD73 (SH3), CD105 (SH2), CD106, CD166, and Stro-1. Moreover they do not express CD45, CD34, CD14 or CD11b, CD79a or CD19 and HLA-DR surface molecules. Mesenchymal stem cells are able to differentiate themselves into osteoblasts, adipocytes and chondroblasts *in vitro* [26]. These cells are defined 'mesenchymal' as they have the ability to maintain the homeostasis of adult mesenchymal tissues [27] because they can develop into more than 200 types of cells [23]. Over the past few years, an important source of adult stem cells has been identified in the dental pulp. The cells in this space match to the criteria of the International Society for Cellular Therapy that define mesenchymal stem cells [28].

In 2006, the scientific world was attracted by a major scientific breakthrough: it was discovered the ability to reprogram somatic cells and bring them back to their pluripotent state through the manipulation of some transcription factors was discovered. These cells were termed "induced pluripotent stem cells" (iPSCs) and they are characterized by the ability to differentiate into the three germ layers just like embryonic stem cells [25]. IPSCs are similar to embryonic stem cells because of their cell morphology, cell-surface markers, telomerase activity, proliferation but there are some differences in gene expression between iPSCs and embryonic stem cells. Nuclear transcriptomes are more complex in embryonic stem cells than in iPSCs [29, 30].

Fetal stem cells, derived from the lifeless bodies of fetuses obtained by spontaneous abortions or still birth, can also be taken from the surgery of ectopic pregnancy. These cells are similar to adult stem cells because they can form a more limited number of cell types [31].

## CONFLICT OF INTEREST

The author confirms that the author has no conflict of interest to declare for this publication.

## ACKNOWLEDEGEMENTS

Declared none.

## REFERENCES

[1]    Basics SC. Stem Cell Information. Bethesda, MD: National Institutes of Health, U.S. Department of Health and Human Services 2009. World Wide Web site

[2]    Saxena AK, Singh D, Gupta J. Role of stem cell research in therapeutic  purpose  hope for new horizon in medical biotechnology. J Exp Ther Oncol 2010; 8(3): 223-33.
       [PMID: 20734921]

[3]    Jones DL, Wagers AJ. No place like home: anatomy and function of the stem cell niche. Nat Rev Mol Cell Biol 2008; 9(1): 11-21.
       [http://dx.doi.org/10.1038/nrm2319] [PMID: 18097443]

[4]    Schofield R. The relationship between the spleen colony-forming cell and the haemopoietic stem cell. Blood Cells 1978; 4(1-2): 7-25.
       [PMID: 747780]

[5]    Dzierzak E, Enver T. Stem cell researchers find their niche. Development 2008; 135(9): 1569-73.
       [http://dx.doi.org/10.1242/dev.019943] [PMID: 18408169]

[6]    Betschinger J, Knoblich JA. Dare to be different: asymmetric cell division in Drosophila, *C. elegans* and vertebrates. Curr Biol 2004; 14(16): R674-85.
       [http://dx.doi.org/10.1016/j.cub.2004.08.017] [PMID: 15324689]

[7]     Clevers H. Stem cells, asymmetric division and cancer. Nat Genet 2005; 37(10): 1027-8.
        [http://dx.doi.org/10.1038/ng1005-1027] [PMID: 16195718]

[8]     Doe CQ, Bowerman B. Asymmetric cell division: fly neuroblast meets worm zygote. Curr Opin Cell
        Biol 2001; 13(1): 68-75.
        [http://dx.doi.org/10.1016/S0955-0674(00)00176-9] [PMID: 11163136]

[9]     Yamashita YM, Fuller MT, Jones DL. Signaling in stem cell niches: lessons from the Drosophila
        germline. J Cell Sci 2005; 118(Pt 4): 665-72.
        [http://dx.doi.org/10.1242/jcs.01680] [PMID: 15701923]

[10]    Morrison SJ, Kimble J. Asymmetric and symmetric stem-cell divisions in development and cancer.
        Nature 2006; 441(7097): 1068-74.
        [http://dx.doi.org/10.1038/nature04956] [PMID: 16810241]

[11]    Mariano ED, Teixeira MJ, Marie SK, Lepski G. Adult stem cells in neural repair: Current options,
        limitations and perspectives. World J Stem Cells 2015; 7(2): 477-82.
        [http://dx.doi.org/10.4252/wjsc.v7.i2.477] [PMID: 25815131]

[12]    Evans MJ, Kaufman MH. Establishment in culture of pluripotential cells from mouse embryos. Nature
        1981; 292(5819): 154-6.
        [http://dx.doi.org/10.1038/292154a0] [PMID: 7242681]

[13]    Martin GR. Isolation of a pluripotent cell line from early mouse embryos cultured in medium
        conditioned by teratocarcinoma stem cells. Proc Natl Acad Sci USA 1981; 78(12): 7634-8.
        [http://dx.doi.org/10.1073/pnas.78.12.7634] [PMID: 6950406]

[14]    Thomson JA, Itskovitz-Eldor J, Shapiro SS, et al. Embryonic stem cell lines derived from human
        blastocysts. Science 1998; 282(5391): 1145-7.
        [http://dx.doi.org/10.1126/science.282.5391.1145] [PMID: 9804556]

[15]    Larijani B, Esfahani EN, Amini P, et al. Stem cell therapy in treatment of different diseases. Acta Med
        Iran 2012; 50(2): 79-96.
        [PMID: 22359076]

[16]    Tondeur S, Assou S, Nadal L. Hamamah s,De Vos J. Biology and potentialities of human embryonic
        stem cells. Ann Biol Clin (Paris) 2008; 66(3): 241-7.
        [PMID: 18558560]

[17]    Skottman H, Hovatta O. Culture conditions for human embryonic stem cells. Reproduction 2006;
        132(5): 691-8.
        [http://dx.doi.org/10.1530/rep.1.01079] [PMID: 17071770]

[18]    Nishikawa S, Jakt LM, Era T. Embryonic stem-cell culture as a tool for developmental cell biology.
        Nat Rev Mol Cell Biol 2007; 8(6): 502-7.
        [http://dx.doi.org/10.1038/nrm2189] [PMID: 17522593]

[19]    Romeo Francesco, Costanzo Francesco, Agostini Massimiliano. Embryonic stem cells and inducible
        pluripotent stem cells: two faces of the same coin? Aging (Albany NY) 2012; 4(12): 878.
        [http://dx.doi.org/10.18632/aging.100513]

[20]    Mitalipov S, Wolf D. Totipotency, pluripotency and nuclear reprogramming. Engineering of stem
        cells. Berlin Heidelberg: Springer 2009; pp. 185-99.

[21]   Wabik A, Jones PH. Switching roles: the functional plasticity of adult tissue stem cells. EMBO J 2015; 34(9): 1164-79.
[http://dx.doi.org/10.15252/embj.201490386] [PMID: 25812989]

[22]   Mimeault M, Batra SK. Concise review: recent advances on the significance of stem cells in tissue regeneration and cancer therapies. Stem Cells 2006; 24(11): 2319-45.
[http://dx.doi.org/10.1634/stemcells.2006-0066] [PMID: 16794264]

[23]   Avasthi S, Srivastava RN, Singh A, Srivastava M. Stem cell: past, present and future--a review article. IJMU 2008; 3(1): 22-31.

[24]   Sousa BR, Parreira RC, Fonseca EA, *et al.* Human adult stem cells from diverse origins: an overview from multiparametric immunophenotyping to clinical applications. Cytometry A 2014; 85(1): 43-77.
[http://dx.doi.org/10.1002/cyto.a.22402] [PMID: 24700575]

[25]   Volarevic V, Ljujic B, Stojkovic P, Lukic A, Arsenijevic N, Stojkovic M. Human stem cell research and regenerative medicine present and future. Br Med Bull 2011; 99(1): 155-68.
[http://dx.doi.org/10.1093/bmb/ldr027] [PMID: 21669982]

[26]   Maleki M, Ghanbarvand F, Reza Behvarz M, Ejtemaei M, Ghadirkhomi E. Comparison of mesenchymal stem cell markers in multiple human adult stem cells. Int J Stem Cells 2014; 7(2): 118-26.
[http://dx.doi.org/10.15283/ijsc.2014.7.2.118] [PMID: 25473449]

[27]   Ma T. Mesenchymal stem cells: From bench to bedside. World J Stem Cells 2010; 2(2): 13-7.
[http://dx.doi.org/10.4252/wjsc.v2.i2.13] [PMID: 21607111]

[28]   Collart-Dutilleul PY, Chaubron F, De Vos J, Cuisinier FJ. Allogenic banking of dental pulp stem cells for innovative therapeutics. World J Stem Cells 2015; 7(7): 1010.

[29]   Brignier AC, Gewirtz AM. Embryonic and adult stem cell therapy. J Allergy Clin Immunol 2010; 125(2) (Suppl. 2): S336-44.
[http://dx.doi.org/10.1016/j.jaci.2009.09.032] [PMID: 20061008]

[30]   Fort A, Yamada D, Hashimoto K, Koseki H, Carninci P. Nuclear transcriptome profiling of induced pluripotent stem cells and embryonic stem cells identify non-coding loci resistant to reprogramming. Cell Cycle 2015; 14(8): 1148-55.
[http://dx.doi.org/10.4161/15384101.2014.988031] [PMID: 25664506]

[31]   Ishii T, Eto K. Fetal stem cell transplantation: Past, present, and future. World J Stem Cells 2014; 6(4): 404-20.
[http://dx.doi.org/10.4252/wjsc.v6.i4.404] [PMID: 25258662]

CHAPTER 2

# Hematopoietic Stem Cells: Identification, Properties and Interest for Clinical Applications

**Nicola Daniele***, **Francesco Zinno** and **Federica Tomassetti**

*Cryolab, Parco Scientifico Università Tor Vergata, Roma, Italia*

**Abstract:** The hematopoietic stem cells (HSCs) are a population responsible of the hematopoiesis's process; they have the characteristic to repeatedly divide or they can mature to generate different cell types, through the process of hematopoiesis. In this regenerative process, the cells are organized in a hierarchical structure: at the summit there are the hematopoietic stem cells and to the base, there is the progeny in differentiation. Hematopoietic cells commissioned to a particular hematic spinneret can be induced to convert themself in cells of the different spinner; another important feature of HSC is plasticity, that is the potential differentiation, thanks to which the cells are capable to undertake phenotypic and functional characteristics of other organs or tissues.

The process of hematopoiesis is regulated by numerous external and internal factors which operate on transcriptional level; this factors can also interact with each other.

Recently, knowledge about HSCs increases more and more; which allows their application also in clinical scope, to permanently treat serious pathologies.

**Keywords:** CD34+, Differentiation, Hematopoietic stem cells, Plasticity, Transplant.

## INTRODUCTION

Hematopoietic stem cells are responsible for the making and turnover of all corpuscular blood's elements; all blood cells originate in fact to pluripotent

---

* **Corresponding author Nicola Daniele:** Cryolab, Parco Scientifico Università Tor Vergata, Roma, Italia; Tel/Fax: +39 039 2109 770; E-mail: Nicola.Daniele@cryolab.solgroup.com

hematopoietic stem cells (PHSC) which constitute 0 0,1% of the nucleated cells in the bone marrow;

Basically, the main characteristics of these cells are:

• the ability to self-renew, or to replicate for many cell cycles;
• non-specialization, as not being mature cells can not carry out any kind of position within the organization;
• may accrue as a result of specific molecular signals and develop themself through a process of differentiation.

Generally these cells are in a quiescent state and remain undifferentiated for indefinite time; in response to various urges they interrupt the quiescence to begin their cycle of mitotic division [1].

Normally, very few HSCs are delegated to the regeneration of hematopoietic cells at any one time; the other remain in the $G_0$ phase of the cell cycle. In fact we know that the blood cells have a limited survival, around 120 days. Their brief existence is linked to the ability to make the most of their task of defense against infection and transport oxygen to tissues; for this reason they must be constantly replaced with new differentiated cells that are fully functioning [2].

This regeneration process takes the name of hematopoiesis and starts from the stem cells present in bone marrow, which divide mitotically and / or differentiate into two classes of multipotent hematopoietic stem cells: the first, the colony forming unit S (CFU- S), will give the cells of the myeloid lineage, including erythrocytes, granulocytes, monocytes and platelets; the second, the colony forming unit- LY (CFU- Ly), is a precursor of the cells of the lymphoid lineage, comprising B and T lymphocytes.

After numerous studies and observations, different antigens have been identified on the outer surface of these cells; on the presence or absence of these molecules, hematopoietic stem cells of man can be divided into two separate classes: cells CD34+ / HLA-DR+ / CD38-, which can divide themselves and became in different type of blood cells, and CD34+ / HLA-DR- / CD38- which can be differentiated in hematopoietic precursors and stromal cells [3].

## Identification of the HSC and Role of the CD34 Antigen

With the aim to study more details of the hematopoietic stem cells, it was necessary to identify and isolate them from all other stem cells and from mature cells. Most of the time the problem that scientists encounter encountered in the identification of these cells, it was the lack of instrumentation. The absence of appropriate methods and technologies to support, have been a major obstacle to the study of this type of cells.

The overcoming of certain limits, in fact, lead to the discovery of the CD34 antigen subsequently; the application of new knowledge about its expression, have proven extremely useful to be able to identify exactly hematopoietic stem cells, in opposition to those already differentiated. In addition the HSC, before maturing, have an increased degree of differentiation and a decrease in the self-renewal capacity, the multipotency and the proliferative potential.

As mentioned earlier, the hematopoietic stem cells express different antigens on their outer surface; among which the most significant is the CD34; it is a trans-membrane glycoprotein which is expressed by the cells only in certain stages of development, when these cells are still immature. The distribution clonal of this antigen varies according to the different moments of cellular development. Consequently, the progenitor cells express the highest levels of CD34, while in cells that are maturing the presence of CD34 receptors on the cell surface decreases drastically [4].

Discoveries and studies in the field of hematopoietic stem cells, were made possible thanks to methods of cell separation using several techniques, including:

- the marking of the cells with monoclonal antibodies (mAb) which, being specific for certain antigens, put in evidence the presence or absence of the receptors on the cell surface, temporarily took under consideration;
- separation with marbles immunnomagnetiche;
- the high-affinity chromatography, based on the interaction between avidin and biotin;
- and finally the flow cytometry.

The last cited technique appeared in the late 60 ' but effectively it became of fundamental importance only in the 80 ', when technological development has produced more and more modern equipment available. Flow cytometry allows to observe the physical and chemical parameters of the cells that are suspended in a certain medium, and today is widely used in the diagnosis of hematological diseases [5].

## Factors of Differentiation

The mechanisms, that induce hematopoietic stem cell to mitotically divide or differentiate into different cell lines and subsequently mature, have not yet been definitively clarified; surely, a central role has the connection between the transcriptional regulation of few genes and the role played by a class of molecules, cytokines, such as the factor of stimulation of hematopoietic colonies.

Some of these molecules are primarily factors that can have a dual role of stimulators or inhibitors of the development, such as interleukin 4 (IL -4); among the factors inhibitors, several cytokines deserve attention because they appear to be effective at very low concentrations. These include interferons, tumor necrosis factor alpha (TNF- alpha), transforming growth factor beta (TGF beta), the inflammatory protein alpha (alpha MIP1) [6].

Many of these cytokines are highly specific for the line of differentiation that the cell choose to mature.

Recent studies have shown that, other cytokines are involved with the function to prevent the process of apoptosis for some cells [7].

## Plasticity of Hematopoietic Stem Cells

Recently, studies on hematopoietic stem cells of the bone marrow have unearth an important feature of these cellelule: plasticity.

HSCs possess a high degree of functional plasticity, which allows them to migrate in some districts of the organism and to generate non- hematopoietic tissues [8]. Subsequently to results of various experiments it has been observed that hematopoietic stem cells taken from the bone marrow of a donor and transplanted into a recipient organism, they can migrate and be found in many organs; more precisely they can mature into:

- on oval cells to generate the liver tissue;
- cardiomyocytes and myocytes respectively in the myocardium tissues in the and skeletal nerve tissue;
- neurons and oligodendrocytes in the central nervous system;
- tubular epithelial cells in the kidney;
- islet beta cells in the pancreas;
- Clara cells in the lung tissue;
- finally, in the cells localized in the skin and in the gastrointestinal tract.

## Clinical Application

Subsequently to a deeper understanding of the role and potential of these cells, recently the focus in this regard is aimed to their use in a clinical setting. Many types of leukemias, lymphomas and autoimmune diseases (Follicular lymphoma, Hodgkin's lymphoma) may be treated to the use of CSE. The bone marrow and the peripheral blood, in addition to umbilical cord blood, are the only sources of immature hematopoietic precursors. In the bone marrow, hematopoietic stem cells comprise approximately 3% - 5%, while in peripheral blood instead 0.03% - 0.05%. Considering the type and stage of disease, and the patient's condition, it is necessary to increase the cell population in order to modulate the therapeutic effects.

To increase the number of circulating hematopoietic progenitors or HSCs in the peripheral blood, one technique called "mobilization" is implemented [9].

This procedure is necessary for to ensure the adequacy and success in case of harvesting of HSCs, also with a single apheresis. Among the elements used to mobilize the HSCs, are included the cytokines administered in presence or absence of chemotherapeutic factors, in the period prior to withdrawal. Experimental results indicate that a more efficient sampling is obtained when the patient is simultaneously subjected to administration of hematopoietic factors and chemotherapeutic agents, at high doses, compared to the two elements used individually. This result highlights a synergistic effect of these two factors in the process of mobilization.

The use of chemotherapeutic and growth factors not only results in a higher release of immature hematopoietic stem cells, but they result in a significant increase of the total number of cells, compared to the conditions at which there isn't this type of treatment.

This allows the collection of a sufficient number of hematopoietic progenitors and their subsequent use in the clinical setting for patients suffering from various diseases, including for example:

• leukemias and lymphomas;
• myelodysplasia;
• myeloma;
• bone marrow aplasia;
• Shortcomings of the bone;
• Congenital disorders such as blood diseases, hemoglobinopathies or serious immunodeficiencies [10].

In case of hematological diseases, in fact, subsequently to cycles of chemotherapy or radiation therapy, the patient is treated with CSE; he undergoes to autologous or allogeneic transplantation, if the hematopoietic stem cells that him receives are belonging to a compatible donor.

For what concerns the autologous transplant, the mobilization is initially induced in the patient by administering of chemotherapy; it is performed a count of CD34+ hematopoietic stem cells by flow cytometry and if there are still malignant cells, it's appropriate an eventual purification.

The measurements of the patient's leukocyte cells are used to determine the appropriate time to start collection procedures. Finally HSCs are preserved and the subsequent cryopreserved at temperature of -196°C in liquid nitrogen. At the time of transplantation, the cells undergo to a rapid thawing and infusion into the patient. Autologous transplantation is primarily indicated for solid tumors and in selected cases of acute myeloid leukemia, acute lymphoblastic leukemia and lymphomas. The use of autologous peripheral blood stem cells is characterized by more rapid recovery of the patient, with possible reduction of infectious and hemorrhagic complications, and a shorter duration of supportive care and hospitalization, compared with bone marrow transplantation.

Allogeneic transplantation is a therapy indicated when the disease cannot be cured with conventional therapies less risky or when these cannot ensure a permanent cure, for example in case of patients affected by pathologies oncohaematological. Allografts from alternative donors such as unrelated individuals or Employing umbilical cord blood must be considered for patients with disease refractory to immunosuppressive therapy. The medullary liquid is directly withdrawal from bones or, thanks to new methods, through apheresis. Preserved in transfusion bags, it's transplanted to the recipient intravenously such as a simple transfusion.

Before to the reinfusion however, the patient must follow cycles of radio - chemotherapy at maximal doses, potentially myeloablative; this is necessary to able to totally eradicate the marrow of the patient and, in the case of malignant diseases, also the residual cells of the disease. This technique takes the name of the "conditioning regimen". The "conditioning regimen" suppresses the marrow of the patient in an irreversible way. The collected cells are infused intravenously in peripheral blood and they having the task to repopulate the bone marrow, producing a new hematopoietic system and immune system. The conditioning regimen is a method also useful to suppress the immunological reactivity of the recipient, so as to avoid the phenomenon of rejection or that of the chronic graft *versus* host disease (GVH) [11].

In allograft, there is an important limitation: the problem of the exact compatibility between donor and recipient. If the HLA system of the patient is not sufficiently equal to the donor, the transplant can become very dangerous for the occurrence of complications. Usually family members have enough an HLA similar to be useful as subjects donors, but there is only 25% of the probability that a relative can be effectively. In the case where this is not possible, there are donor's banks to which may be required to find a Matched Unrelated Donor (MUD). Besides to the identity between HLA systems, the success of the transplant between non-consanguineous, also depends on other variables to consider, such as the type of the disease, the stage of development of pathology, conditions of the patient [12].

It happiness because the patient, in the weeks before allograft, should be subjected to extremely aggressive therapies. That therapies expose the patience to infections because of immune deficiency; subsequently to transplantation instead, many months are necessary for the recovery of normal immune functions and avoid to complications such as GVHD, immunodeficiency longest expected and/or recurrence [13].

As regards the application of hematopoietic stem cells in a clinical setting, it's important remember that a very important role has recently been attributed to the blood from the umbilical cord. In recent years, in fact, it has become increasingly common practice to retain the newborn's umbilical cord, immediately after birth. Several banks appointed to the cord blood preservation have emerged in recent years; blood containing hematopoietic stem cells is thus stored and the storage of stem cells can occur for many years.

The retention periods are so long term that if ever the child will not have necessity for himself, the cells could be used also by another member of the family, in the case the subject is compatible to transplantation. In fact it was discovered that, in opposition to which happens in allogeneic transplants with stem cells taken from circulating blood or bone marrow where there must be a very good compatibility between recipient and donor, the use of blood belonging to the umbilical cord does not need the same high compatibility.

Most recent studies have allowed treatments with the use of stem cells from the umbilical cord; this discovered are innovative and research is in continually movement to improve the applications and find new potentiality of stem cells.

Now, some diseases are not yet treatable through the use of hematopoietic stem cells but in the near future probably, new discoveries will extend the application not only to those currently treatable diseases, but also to those that are still totally unknown to our.

## CONFLICT OF INTEREST

The authors confirm that the author have no conflict of interest to declare for this publication.

## ACKNOWLEDEGEMENTS

Declared none.

## REFERENCES

[1]     Gartner LP, Hiatt JL. Istologia. Edises 2005; 30: 220-50.

[2]     Dexter M, Allen T. Haematopoiesis. Multi-talented stem cells? Nature 1992; 360(6406): 709-10.
        [http://dx.doi.org/10.1038/360709a0] [PMID: 1465141]

[3]     Jansen J, Hanks S, Thompson JM, Dugan MJ, Akard LP. Transplantation of hematopoietic stem cells
        from the peripheral blood. J Cell Mol Med 2005; 9(1): 37-50.
        [http://dx.doi.org/10.1111/j.1582-4934.2005.tb00335.x] [PMID: 15784163]

[4]     Satterthwaite AB, Borson R, Tenen DG. Regulation of the gene for CD34, a human hematopoietic
        stem cell antigen, in KG-1 cells. Blood 1990; 75(12): 2299-304.
        [PMID: 1693527]

[5]     He Qi, Chang Chun-Kang, Xu Feng, Zhang Qing-Xia, Shi Wen-Hui, Li Xiao. Purification of bone
        marrow clonal cells from patients with myelodysplastic syndrome *via* IGF-IR. Plos One 2015; 10: 4-
        10.

[6]     Katayama N, Clark SC, Ogawa M. Growth factor requirement for survival in cell-cycle dormancy of
        primitive murine lymphohematopoietic progenitors. Blood 1993; 81(3): 610-6.
        [PMID: 7678992]

[7]     Peschel C, Paul WE, Ohara J, Green I. Effects of B cell stimulatory factor-1/interleukin 4 on
        hematopoietic progenitor cells. Blood 1987; 70(1): 254-63.
        [PMID: 3496134]

[8]     Jiang Y, Jahagirdar BN, Reinhardt RL, *et al.* Pluripotency of mesenchymal stem cells derived from
        adult marrow. Nature 2002; 418(6893): 41-9.
        [http://dx.doi.org/10.1038/nature00870] [PMID: 12077603]

[9]     Kroger N, Regens H, Kruger W, Loliger C, Carrero I. A randomized comparison of once *versus* twice
        daily recombinant human granulocyte colony-stimulating factor for stem cell mobilization in healthy
        donors for allogenic transaplantation. Br J Haematol 2000; 111: 761-5.

[10]    Pettengell R, Testa NG, Swindell R, Crowther D, Dexter TM. Transplantation potential of
        hematopoietic cells released into the circulation during routine chemotherapy for non-Hodgkins
        lymphoma. Blood 1993; 82(7): 2239-48.
        [PMID: 7691253]

[11]    Kessinger A, Armitage JO. The evolving role of autologous transplantation following high-dose
        therapies for malignancies. Blood 1991; 77: 211-3.

[PMID: 1670758]

[12]    Huang S, Terstappen LW. Lymphoid and myeloid differentiation of single human CD34+, HLA-DR+, CD38- hematopoietic stem cells. Blood 1994; 83(6): 1515-26.
[PMID: 7510144]

[13]    Yamazaki H. Hematopoietic stem cell transplantation for acquired aplastic anemia. Rinsho Ketsueki 2015; 56: 2153-9.

# CHAPTER 3

# Stem Cell Therapy Applications: The Challenge of Regenerative Medicine

**Fulvia Fraticelli***

*Cryolab, University of Rome "Tor Vergata", Rome, Italy*

**Abstract:** Regenerative medicine is a new concept of developing medicine. This field of science concerns reconstruction and repair of damaged tissues and organs. In physiological conditions, their integrity and functionality are ensured by the presence of adult stem cells that maintain and renew effectively all cell types. When there are injuries resulting from different causes, its essential to reconstitute the original structure and this represents the goal of regenerative medicine. The aim is to employ stem cells in clinic as cell therapy. For obtaining this result, its necessary to know the biology of stem cells and then its essential to have technologies useful to regenerate them in culture and allow their vitality. Today stem cells are a therapeutic reality: their potentiality helped in healing many kinds of diseases and restore the health of patients. With the progress of research and the development of new therapy protocols, it will be possible to expand stem cell therapy to different specialized areas of pathology.

**Keywords:** Cell therapy, Hematopoietic stem cells, Mesenchymal stem cells, Regenerative medicine, Stem cells.

## INTRODUCTION

The discovery of stem cells has been one of the greatest revolutions in medical field because, thanks to their typical characteristics, they are used currently in clinical practice. From the earliest research, the extraordinary ability of these cells to self-renew and at the same time, differentiate themselves into unlimited cell types has been noticed. This plasticity has been employed in therapy for treatment

---

* **Corresponding author Fulvia Fraticelli:** Cryolab, University of Rome "Tor Vergata", Rome, Italy; Tel: +39 039 2109770; Fax: +39 039 2109765; E-mail: fulvia.fraticelli@gmail.com

care and complete resolution, sometimes, of several diseases. The advancement of knowledge, the multidisciplinary study of pathology and the availability of new methodologies for investigation of cells contributed to a radical change in medicine. Until a few years ago, diseases were treated with traditional approaches, essentially the pharmacologic ones. But, in recent times, a new field of medicine has developed: the field of regenerative medicine. The bone marrow contains mainly two different kinds of cell: hematopoietic stem cells (HSCs) and stromal ones, also called mesenchymal stem cells (MSCs).

Hematopoietic stem cells have an undeniable therapeutic value, because they are able to repair and regenerate also non hematopoietic tissues, so their clinical use concerns different fields of pathology. The main use of HSCs in therapy consists in transplantations, although current studies are proving that they are applied in heart diseases, in bone and cartilage alterations, cancer, neurodegenerative impairment and so on [1].

The model of stem cells is the hematopoietic type, derived from bone marrow, peripheral blood and umbilical cord. But also mesenchymal stem cells are important in therapy: they have different origins but, according to several properties of renewal and differentiation, they can be used in cell therapy and regenerative medicine in a specific way [2].

MSCs have a great potentiality in clinic, because they differentiate in several populations, like vascular, endothelial, muscular, and non-mesodermal ones. So, they can be used for tissue repair especially thanks to secretion of growth factors and cytokines. They should be able to work as inhibitors of cancer proliferation, switching off inflammatory reactions, checking immune responses and so reestablishing integrity of tissues [3].

In addition to conventional medicine, nowadays it is trying to extend the application of stem cells as replacement therapies for several different diseases. In particular, the use of induced pluripotent stem cells (IPS) seems to be promising. They are taken from patients and are capable of differentiating into various cells, applying the technology of reprogramming somatic cells. Using cells coming from pluripotent stem cells, its possible to realize a cell therapy for many different body systems. Availability of a stock of specific cell types is the most important condition for regenerative medicine to perform and so realizing specific therapies for individual pathologies [4].

The basic requirement for application of stem cells in clinic consists in the employment of substrates on which growth, regeneration and self-renewal are ensured. Thanks to these materials, its possible to have many types of substances for maintenance of human pluripotent stem cells. In this way, regenerative medicine will be realized and a stem cell based therapy will be achieved. A large variety of materials exist: the main characteristics being the promotion of self-renewal, pluripotency and expansion. The recommendable substrate for clinical use would be inexpensive, easy to manipulate, rapid to use, safe and easily sterilized. Synthetic polymers, synthetic peptides, hydrogels and extracellular matrix proteins are some of the substrates currently available for an optimal clinical appliance [5].

## Hematopoietic Stem Cells in Therapy

Hematopoietic stem cells are very useful in therapy, since they are able to regenerate and reproduce many specific cells. Their expansions in culture allow to obtain easily a large amount of materials, in particular blood cells for hematological disorders. As a result of new techniques, it was possible to create HSCs from embryonic stem cells (ESCs) and IPS. The natural growth, development and survival of these cells must be reproduced *in vitro* for making best use of HSCs in clinic. The most important application includes transplants and cell-based therapies [6].

Hematopoietic stem cells transplantation represents the gold standard treatment for blood diseases. Thanks to the regeneration carried out by stem cells, its possible to achieve engraftment and survival, which are essential for the success of transplants. Since there is a network of signals which cooperate to ensure the functionality of cells within stem cells niche, the comprehension of multiple interactions is primary for using cells in therapy. By analyzing specific pathways and using *in vitro* technologies, the success of transplantation using HSCs from ESCs and IPS can be increased. In this way, some of the problems related to intrinsic limits of cells could be solved with current strategies [7].

Hematopoietic stem cells are very complex and heterogeneous cells and so they are capable of regenerating and differentiating into various lineages with a specific mechanism. According to this peculiarity, they are able to restore impaired tissues and are used for patient's care. It has been shown that HSCs have different capacities of differentiation and renewal. These mechanisms can be used in regenerative medicine in a different way, depending on the purpose. In practice, it might expand the most suitable subset for clinical application and so its possible to obtain specific cells for every tissue [8].

**Mesenchymal Stem Cells in Therapy**

Mesenchymal stem cells are used as therapy in many patients for acute and chronic diseases. Their application and modality of action is relevant for clinical use. Thanks to the strategy of culture, its possible to expand MSCs and so obtain a wide range of products.

In order to use mesenchymal cells as therapeutic tool, it is necessary to identify and understand nature and mechanisms of these cells inside the bone marrow and tissues. Besides, a rapid and secure culture's strategy should be available to demonstrate the potential of stem cells during clinical trials. After performing studies, it will be possible to use these cells safely [9].

This method would be important for the treatment of various diseases and maintenance of general homeostasis of cells and their capacity of differentiation into several cell types.

They are isolated from different tissues and are able to expand fastly in multiple lineages. Their clinical application is about immune and non-immune disorders, since they can produce adipocytes, chondrocytes, myoblasts, osteoblasts, and neuron-like cells [10].

MSCs are ideal cells for therapeutic approach: they are easy to use and obtain, they grow in a short time and their results in clinic are safe and promising. According to the important studies, mesenchymal cells are useful for regenerative medicine and inflammatory diseases. They operate at various levels, ensuring the protection of cells from damaging event and at the same time promoting tissue repair [11].

## Therapeutic Applications

As mentioned earlier, "regenerative medicine" means that human cells are able to be used in therapy for treatment of a number of medical conditions. There are several diseases that require the restore of function and the return to normality. Thanks to the technological approaches, regenerative medicine allows to replace lost or injured cells, tissues and organs.

Currently, there is a broad spectrum of application of regenerative medicine: different areas of specialized medicine commonly use these methods and strategy [12].

In this chapter, some examples of therapeutic applications are cited and a table shows the main clinical studies regarding the use of stem cells in the world (Table 1).

Table 1. Clinical trials of stem cells.

| No | Title | Sponsors | Study Type | Study Phase | Recruitment Status | Conditions | Estimated Completion Date |
|---|---|---|---|---|---|---|---|
| 1 | Clinical Trial of Stem Cell Based Tissue Engineered Laryngeal Implants | University College, London | Interventional | Phase 1, Phase 2 | Recruiting | Disorder of Upper Respiratory System; Laryngostenosis; Tracheal Stenosis | December 2018 |
| 2 | Allogeneic Heart Stem Cells to Achieve Myocardial Regeneration | Capricor Inc. | Interventional | Phase 1, Phase 2 | Recruiting | Myocardial Infarction | September 2021 |

*(Table 1) contd.....*

| No | Title | Sponsors | Study Type | Study Phase | Recruitment Status | Conditions | Estimated Completion Date |
|---|---|---|---|---|---|---|---|
| 3 | Periodontal Tissue Regeneration Using Autologous Periodontal Ligament Stem Cells | Fourth Military Medical University | Interventional | Phase 1 | Recruiting | Periodontal Pocket | December 2014 |
| 4 | Autologous Mesenchymal Stem Cells Transplantation in Women With Premature Ovarian Failure | Sayed Bakry, Al-Azhar University | Interventional | Phase 1, Phase 2 | Recruiting | Premature Ovarian Failure | November 2016 |
| 5 | Intravenous Injection of Adipose Derived Mesenchymal Stem Cell for ALS | Royan Institute | Interventional | Phase 1 | Recruiting | Amyotrophic Lateral Sclerosis | May 2016 |
| 6 | Allogeneic Mesenchymal Stem Cell Transplantation in Tibial Closed Diaphyseal Fractures | Royan Institute | Interventional | Phase 2 | Recruiting | Tibial Fracture | April 2016 |
| 7 | Safety and Efficacy Studies of Umbilical Mesenchymal Stem Cell for Liver Cirrhosis | Alliancells Bioscience Corporation Limited | Interventional | Phase 1, Phase 2 | Recruiting | Liver Cirrhosis | October 2015 |
| 8 | Allogenic AD-MSC Transplantation in Idiopathic Nephrotic Syndrome | Royan Institute | Interventional | Phase 1 | Recruiting | Focal Segmental Glomerulosclerosis | October 2017 |

*(Table 1) contd.....*

| No | Title | Sponsors | Study Type | Study Phase | Recruitment Status | Conditions | Estimated Completion Date |
|----|-------|----------|------------|-------------|--------------------|------------|---------------------------|
| 9 | Blood Derived Autologous Angiogenic Cell Precursor Therapy in Patients With Critical Limb Ischemia (ACP-CLI) | Hemostemix | Interventional | Phase 2 | Recruiting | Critical Limb Ischemia | August 2017 |
| 10 | Muscle Progenitor Cell Therapy for Urinary Incontinence | Gopal Badlani, MD, Wake Forest School of Medicine | Interventional | Phase 1, Phase 2 | Recruiting | Urinary Incontinence | December 2017 |
| 11 | Combination of Mesenchymal and C-kit+ Cardiac Stem Cells as Regenerative Therapy for Heart Failure | The University of Texas Health Science Center, Houston | Interventional | Phase 2 | Recruiting | Ischemic Cardiomyopathy | August 2018 |
| 12 | Safety Study of Liver Regeneration Therapy Using Cultured Autologous BMSCs | Yamaguchi University Hospital | Interventional | Phase 1 | Recruiting | Liver Regeneration | March 2017 |
| 13 | Adipose Derived Regenerative Cellular Therapy of Chronic Wounds | Tower Oupatient Surgical Center | Interventional | Phase 2 | Recruiting | Diabetic Foot, Venous Ulcer, Pressure Ulcer | September 2015 |
| 14 | Mesenchymal Stem Cell in Patients With Acute Severe Respiratory Failure | Asan Medical Center | Interventional | Phase 2 | Recruiting | Respiratory Distress Syndrome, Adult | December 2016 |

These studies are provided by U.S. National Institutes of Health and are available on the website: ClinicalTrials.gov (http://clinicaltrials.gov)

- Cancer: it is desirable, in patients with cancer, restoring hematopoiesis after chemotherapy. So, the main strategy is the transplantation of stem cells. But there are several limits and today it is possible to make use of induced pluripotent stem cells or, alternatively, acting on stem cell niche in order to support endogenous repair [13].

- Regeneration of tissues: the potentiality of stem cells about generation of differentiated lineages ensures their application to reconstitute injured tissues. The regulation of proliferation and differentiation are essential for obtaining a complete repopulation after damage. It is possible, thanks to plasticity and mobilisation of stem cells, to get the restoration of the affected area [14].

- Heart diseases: cardiovascular injuries represent one of the main problems for medicine, because therapeutic strategies are restricted. The aim of regenerative therapy is linked with reproduction of cardyomyocites through stem cells. Using fibroblasts, infact, cardiomyocytes are generated and they can replace infarcted area or failing heart [15].

- Neurologic disorders: there are many different diseases, which affect the nervous system. In each of them, the nerve cells involved are not the same. Some types are selectively present in certain disorders as compared to other. It is possible to have neural human cells through differentiation of stem cells and lineage conversion from somatic ones [16].

- Bones: skeletal diseases may have a genetic and non-genetic origin. When there is an alteration of functionality and a defect in turnover of the matrix occurs, disequilibrium of homeostasis is responsible for bone diseases. Stem cells and stromal ones can contribute to restore functionality and to form a healthy tissue [17, 18].

- Liver diseases: the liver is one of the most important organs that govern many functions inside the body. It is characterized by a singular capacity of regeneration and, in combination with stem cells, it is possible to treat and solve hepatic failure [19].

- Corneal diseases: regenerative medicine for the cornea consists in the use of cultured stem cells coming from oral mucosal epithelial cells. Infact, they have characteristics common to that of corneal epithelium. Despite this technique has not yet been approved, it proves to be very interesting as an alternative to keratoplasty [20].

- Renal diseases: kidney damage manifests itself in various ways, so treatments are different and are designed for restoring functionality. There are many approaches based on stem cells. Thanks to their easy manipulation, the cell therapy method ensures kidney regeneration after injury. The sources of cells are several: their integration with tissue damage determines growth and restart of kidney functions [21].

- Lung damage: when lung is injured, especially by chronic diseases, therapies with stem cells can improve patient's health. In order to simplify repair, regenerative medicine makes use of HSCs and MSCs from various sources. In addition to cell therapy, then, the method of tissue bioengineering is also available. In association with stem cells-based therapy, it will be possible, in the future, to work in a synergic way to obtain promising results [22].

## CONCLUDING REMARKS

In recent years, progresses in science have contributed to an expansion of knowledge on the mechanisms of diseases etiopathology. Through the study of relationship among cells in our body, it was possible to identify new targets for research and resolution of illnesses. Treatment of pathologies and care of many clinical cases represent the main challenge of actual medicine. Thanks to the support of technology and the discovery of more advanced techniques, regenerative medicine is the most important technique of developing medicine. Regenerative medicine has developed recently with advances in biotechnology and tissue engineering. Only by implementing research and cooperating with other medical fields, it will be possible to achieve, in the future, a certain understanding of pathological mechanisms.

## CONFLICT OF INTEREST

The author confirms that author has no conflict of interest to declare for this publication.

## ACKNOWLEDGEMENTS

Declared none.

## REFERENCES

[1]     Porada CD, Atala AJ, Almeida-Porada G. The hematopoietic system in the context of regenerative medicine. Methods 2016; 99: 44-61.
[http://dx.doi.org/10.1016/j.ymeth.2015.08.015] [PMID: 26319943]

[2]     Trivanović D, Jauković A, Popović B, *et al.* Mesenchymal stem cells of different origin: Comparative evaluation of proliferative capacity, telomere length and pluripotency marker expression. Life Sci 2015; 141: 61-73.
[http://dx.doi.org/10.1016/j.lfs.2015.09.019] [PMID: 26408916]

[3]     Stoltz JF, de Isla N, Li YP, Bensoussan D, Zhang L. Stem cells and regenerative medicine: Myth or reality of the 21th century. Stem Cells Int 2015; 2015(3): 734731.

[4]     Fox IJ, Daley GQ, Goldman SA, Huard J, Kamp TJ, Trucco M. Use of differentiated pluripotent stem cells in replacement therapy for treating disease. Science 2014; 345(6199): 1247391.
[http://dx.doi.org/10.1126/science.1247391] [PMID: 25146295]

[5]     Enam S, Jin S. Substrates for clinical applicability of stem cells. World J Stem Cells 2015; 7(2): 243-52.
[http://dx.doi.org/10.4252/wjsc.v7.i2.243]

[6]     Nakajima-Takagi Y, Osawa M, Iwama A. Manipulation of hematopoietic stem cells for regenerative medicine. Anat Rec (Hoboken) 2014; 297(1): 111-20.
[http://dx.doi.org/10.1002/ar.22804] [PMID: 24293004]

[7]     Van Zant G, Liang Y. Concise review: hematopoietic stem cell aging, life span, and transplantation. Stem Cells Transl Med 2012; 1(9): 651-7.
[http://dx.doi.org/10.5966/sctm.2012-0033] [PMID: 23197871]

[8]     Muller-Sieburg CE, Sieburg HB, Bernitz JM, Cattarossi G. Stem cell heterogeneity: implications for aging and regenerative medicine. Blood 2012; 119(17): 3900-7.
[http://dx.doi.org/10.1182/blood-2011-12-376749] [PMID: 22408258]

[9]     Griffin MD, Elliman SJ, Cahill E, English K, Ceredig R, Ritter T. Concise review: adult mesenchymal stromal cell therapy for inflammatory diseases: how well are we joining the dots? Stem Cells 2013; 31(10): 2033-41.

[10]    Ren G, Chen X, Dong F, *et al.* Concise review: mesenchymal stem cells and translational medicine: emerging issues. Stem Cells Transl Med 2012; 1(1): 51-8.
[http://dx.doi.org/10.5966/sctm.2011-0019] [PMID: 23197640]

[11]    Parekkadan Biju, Milwid Jack M. Mesenchymal stem cells as therapeutics. Annu Rev Biomed Eng 2010; 12: 87-117.
[http://dx.doi.org/10.1146/annurev-bioeng-070909-105309]

[12]    Mason C, Dunnill P. A brief definition of regenerative medicine 2008. Fut Med Ltd ISSN 1746-0751.
[http://dx.doi.org/10.2217/17460751.3.1.1]

[13]  Lane SW, Williams DA, Watt FM. Modulating the stem cell niche for tissue regeneration. Nat Biotechnol 2014; 32(8): 795-803.

[14]  Wabik A, Jones PH. Switching roles: the functional plasticity of adult tissue stem cells. EMBO J 2015; 34(9): 1164-79.
[http://dx.doi.org/10.15252/embj.201490386] [PMID: 25812989]

[15]  Yamakawa H, Ieda M. Strategies for heart regeneration: approaches ranging from induced pluripotent stem cells to direct cardiac reprogramming. Int Heart J 2015; 56(1): 1-5.
[http://dx.doi.org/10.1536/ihj.14-344]

[16]  Ichida JK, Kiskinis E. Probing disorders of the nervous system using reprogramming approaches. EMBO J 2015; 34(11): 1456-77.
[http://dx.doi.org/10.15252/embj.201591267] [PMID: 25925386]

[17]  Bianco P. Stem cells and bone: A historical perspective Published Online. 2014.

[18]  Riminucci M, Remoli C, Robey PG, Bianco P. Stem cells and bone diseases: New tools, new perspective Published Online. August 26, 2014. Edited by: Frank P. Luyten

[19]  Bhatia SN, Underhill GH, Zaret KS, Fox IJ. Cell and tissue engineering for liver disease. Sci Trans Med 2014; 6(245): 245sr2.

[20]  Oie Yoshinori, Nishida Kohji. Regenerative medicine for the cornea. Biomed Res Int 2013.
[http://dx.doi.org/10.1155/2013/428247]

[21]  Chung HC, Ko IK, Atala A, Yoo JJ. Cell-based therapy for kidney disease. J Urol 2015; 56(6): 412-21.
[http://dx.doi.org/10.4111/kju.2015.56.6.412]

[22]  Yang J, Jia Z. Cell-based therapy in lung regenerative medicine. Regen Med Res 2014; 2(1): 7.
[http://dx.doi.org/10.1186/2050-490X-2-7] [PMID: 25984335]

# CHAPTER 4

# Pluripotent Stem Cells: Basic Biology and Translational Medicine

**Filippo Zambelli[1,2], Lucia De Santis[3] and Rita Vassena[4,\*]**

[1] *Vrije Universiteit Brussel, Research Group Reproduction and Genetics, Brussels, Belgium*

[2] *S.I.S.Me.R. Reproductive Medicine Unit, Bologna, Italy*

[3] *San Raffaele Scientific Institute, Vita-Salute University, Dept Ob/Gyn, IVF Unit, Milan Italy*

[4] *Clinica EUGIN, Barcelona, Spain*

**Abstract:** The derivation of human embryonic stem cells in the last decades, made possible by the parallel and growing development of *in vitro* fertilization and embryo cryopreservation technologies, have opened the door for regenerative medicine. The study of cell replacement in loss of function diseases has received further impulse by the derivation of induced pluripotent cells less than 10 years ago. Currently, pluripotent cells are extensively employed in disease modeling, toxicology testing, and drug discovery. Phase I clinical trials with both embryonic and induced pluripotent cells derivates have been underway for a few years now, and initial results have been published recently. As the field of regenerative medicine moves forward at an impressive pace, we aim to review the origin and characteristics of the different kind of pluripotent stem cells, their potential use in key translational areas, and the challenges and opportunities that we face for their integrated use in a modern and personalized medicine.

**Keywords:** Cell therapy, Differentiation, *In vitro* disease modeling, *In vitro* drug screening, *In vitro* embryo culture, Induced pluripotency, Pluripotent stem cells, Regenerative medicine.

---

\* **Corresponding author Rita Vassena:** Clinica EUGIN, Barcelona, Spain; Tel: +34-93 322 11 22; E-mail: rvassena@eugin.es

## THE CONCEPT OF PLURIPOTENCY

A chapter on the promises of translational medicine in the field of stem cell biology, promises brought about by our ability to harness pluripotent cell, would not be starting off on the right foot without a brief definition of what pluripotency is.

Etymologically, the term pluripotency derives from the union of the Latin words "*totus*" which means entirely or all, and "*potens*", which means having ability or power. Something pluripotent is therefore something with the ability of being, or becoming, all. When applied to cell biology, the term pluripotent acquires a more specific meaning and describes the capability of a cell to *divide indefinitely **while** maintaining the ability to differentiate into all cell types of an organism*. In this "while" resides the unique power of pluripotent cells. In biology, in fact, we can find several examples of cells that are able to divide indefinitely, from cancer cell lines, to primary lines immortalized by viral transduction. In the same way, there are examples of cells which are capable of differentiating in several cell types, like for instance the progenitors of an organ or tissue, or even into all the cell types that form an individual, like the fertilized oocyte or zygote. However, these states of pluripotency are not sustained over time, and in fact are present only for a few hours/cell divisions in the life of an organism. When left to their own devices, naturally occurring pluripotent states tend to differentiate quickly to a more differentiated, less potent state.

How is pluripotency tested? From its definition, it is clear that the ideal pluripotent cell should be able to give rise to a complete individual, however a series of proxy tests for differentiation potential have been devised. The most stringent proof of pluripotency, so far only available in mice, is given by the tetraploid complementation assay [1]. This test is based on the observation that when tetraploid (4n) and diploid (2n) cells are mixed together in a preimplantation mouse embryo, the tetraploid cells will segregate to the trophectoderm, leaving the diploid ones to form the inner cell mass (ICM). It is possible to create tetraploid early embryos by fusion of the two blastomeres of a 2-cell embryo. Diploid cells to be tested for pluripotency are added to the 4n cells, and a chimeric embryo is reconstructed. Because of the selective segregation of 4n cells, the resulting pup will be formed entirely from the cells to be tested, and if a fertile individual is born, pluripotency is proven with the highest degree of confidence. A less stringent variation of the tetraploid supplementation test is the simple chimera formation test. A chimera is an individual which is made up by more than one genetically distinct cells. When injecting pluripotent cells into the blastocele of a developing embryo, some of them will be incorporated into the ICM, and will give rise to part of the developing individual. The degree and quality of this integration is usually taken as an indication of the level of pluripotency. Interspecies chimeras between human pluripotent cells and mice embryos were not produced until a few years ago, but are a technical possibility. Nonetheless, due to both legal and ethical concerns, the most common and accepted proof of pluripotency in human cells is the teratoma test. Teratomas are benign tumors which may be composed by derivatives of the three embryonic layers; teratomas are usually found in newborn or developing feti, but can be also discovered accidentally in adult individuals. In stem cell research, a teratoma is produced, intentionally, by injecting a pellet of cells of putative pluripotent potential into an immunologically suppressed host, usually a SCID mouse. Injection sites can vary, from subcutaneous to intratesticular to sub capsular in kidney and liver. Usually, if the tested cells are indeed pluripotent, a teratoma will form within 8 to 10 weeks. A less stringent *in vitro* test of pluripotency is through the production of so called embryoid bodies (EBs). EBs are produced when pluripotent cells are removed from an undifferentiated milieu, and allowed to aggregate and grow, usually in suspension. Pluripotent cells in this condition tend to form aggregates that differentiate in an undirected way, and presenting, often, cell populations which are precursors of more differentiated cell types and tissues. Cells that are not pluripotent, but that can still differentiate into some cell types (called multipotent) can also differentiate to a certain degree in EBs, thus forming EBs alone is not considered a stringent test of pluripotency.

Regardless of how pluripotency is tested, it takes both a cell population able to differentiate into all cell types of an organism, as well as an *in vitro* system which is able to override the differentiation signals to maintain a pluripotent cell population. While such an efficient *in vitro* culture system for human cells has been developed in the last 20 years and is still under improvement in the present days, the main cell types that can be expected to be pluripotent have been identified, and will be described further along the chapter.

## THE CONTRIBUTION OF ASSISTED REPRODUCTION TO THE PLURIPOTENT STEM CELL FIELD

The development of the pluripotent stem cell field, both in basic science and translational medicine, has been made possible by the previous development of assisted reproduction technologies, and in particular by *in vitro* culture systems and reliable cryopreservation protocols for human embryos. Since Louise Brown's birth the number of *in vitro* fertilization (IVF) treatments has exponentially increased worldwide. In a typical IVF treatment, not all embryos that develop *in vitro* following fertilization can be replaced in the uterine cavity at the same time, since this would likely result in a higher order pregnancy with significant health risk for both the mother and the future children. Embryo cryopreservation is a routine strategy deployed during assisted reproduction treatments to store surplus embryos for future efforts, with the final aim to increase the cumulative live birth rate of one ovarian stimulation cycle. Due to IVF own success, some cryopreserved embryos will ultimately not be transferred to the uterus of the patient, for instance because she would have already conceived during a previous transfer, or after the couple have exhausted their reproductive wishes.

The importance of the contribution of human embryology to embryonic stem cell biology and derivation further appears when considering that, although the studies on mouse stem cells have provided important insights into methodologies and more deep knowledge of stem cell biology, the two species present different biochemical and *in vitro* culture requirements [2].

## PLURIPOTENT CELL TYPES

### EC Cells

One of the first pluripotent cell populations to be cultured and expanded *in vitro* were cells derived from teratocarcinomas. Teratocarcinomas are a kind of malignant germ cell tumor. Although these tumors have been known for decades, it was not until the mid sixties than embryonic carcinoma (EC) cells could be established *in vitro* [3], where they showed both the ability to grow indefinitely and to differentiate into derivatives of the three germ layers. Besides being clearly pluripotent, EC cells were the first cancer cells which have been demonstrated to be able to generate a tumor from their undifferentiated state. In fact, their clonogenic capacity, *i.e.* the ability of one single EC cell to re-grow a tumor *in vivo*, demonstrated for the first time the so called "stem cell theory of cancer", is still extensively studied to date, decades later.

## EG Cells

Embryonic germline (EG) cells are derived when primordial germ cells (PGC) which resides in the developing gonads are cultured *in vitro*. In contrast to EC cells, EG cells were derived only relatively recently in both mice [4] and human [5]. EG cells do present a series of cellular attributes that are common to other pluripotent cells, for instance they are very similar morphologically, and express some of the core regulators of pluripotency such as Oct4. However, they also maintain some characteristics of the germline cells from which they derive, especially at the epigenetic level, including X chromosome activation.

The derivation and especially the expansion and maintenance *in vitro* of human EG cells is comparatively difficult, as these cells tend to differentiate easily *in vitro* and not to fare well in long term culture. Moreover, the formation of teratomas, which as mentioned is a hallmark of pluripotency for human cells, has not been proven so far. EG cells in the human species are currently rarely studied due to difficulties in procurement, derivation, and long term culture.

## hESC

Human embryonic stem cells have been derived for the first time in 1998 [6] in a seminal experiment by which human blastocysts have been cultured on a feeder layer of fibroblasts with a medium containing human serum. Although current culturing technique dispense of human serum in favor of more defined substitutes, and employ TGFb as the main cytokine, this technique has allowed for the growth of the cells from the inner cell mass (ICM), which usually give rise to the embryo proper during development, beyond a few division, forming flat colonies of tightly growing cells with a high nucleus to cytoplasm ratio. The colonies could be expanded indefinitely *in vitro* and, when the undifferentiating environment *in vitro* was removed, were able to differentiate to derivatives of the 3 germ layers (endoderm, ectoderm and mesoderm) indicating their potential to give rise to all tissues of an organism.

Although hESC lines are by and large derived from blastocyst stage embryos, they have been also derived from earlier stages such as morula [7], 4 cells [8], from single blastomeres [9], and from monoparental constructs such as parthenotes [10]. hESC can give rise to derivatives of the 3 germ layers in teratomas, but are unable to integrate into mouse embryos to form chimeras at their commonly found state, also called "primed" state.

Although some countries, such as the USA and Belgium, allow for the creation for embryos with the purpose of using them in stem cell research, currently hESC are mostly obtained from embryos that have been donated by couples undergoing assisted reproduction, and who have fulfilled their reproductive desires. The embryos are then donated to research, where possible and if the couple wishes so. Another, less frequent source of embryos for hESC derivation are embryos considered of low quality, and that are discarded during an IVF treatment, since they are too compromised to either implant or withstand cryopreservation. Finally, hESC can be derived from human embryos which have been diagnosed with a severe disease following preimplantation genetic diagnosis. These embryos are not available for transfer into the uterus of the recipient woman, since they would give rise to a fetus affected from the disease that the procedure is attempting to prevent, and are usually discarded. The derivation of hESC from these embryos is particularly interesting, since the cells will all be carrying the diseased genes or chromosomes, and upon differentiation, they will very likely present a phenotype very similar to that found in the affected individuals [11, 12]. These cell lines serve therefore as a "disease in a Petri dish" model, and have allowed the study of hard to access cell population in diseased individuals, with the aim to foster the search for therapies [13 - 15].

# iPS

iPS is the acronym for induced pluripotent stem cells, which was coined by the Japanese researcher Shinya Yamanaka in a seminal paper published in 2006 [16]. In a series of experiments which have all but reshaped the history of stem cell biology, and which were the basis of the motivation for the Nobel Prize for Physiology or Medicine awarded to him and Sir John Gurdon in 2012, he showed that a completely differentiated cells could be "made" to behave by and large like embryonic stem cells by the forced over expression of a limited number of exogenous transcription factors. In the original set of studies, mouse and human fibroblasts were transfected with a cocktail of more than 20 different transcription factors, all considered relevant to continuous division, early embryonic development, or stem cell maintenance [16, 17]. Then, each one of the transcription factors was removed sequentially from the mix, and the effects on the transfected cells were analyzed. Following this painstaking work, the investigators concluded that forced over expression of 4 exogenous transcripts: Oct4, Sox2, Klf4, and c-Myc was sufficient to change the epigenetic landscape of the fibroblasts, and to convert them in bona fide pluripotent cells. Although the presence of c-Myc to the mix rendered the conversion more efficient, only Oct4, Sox2, and Klf4 were the only strictly required transcripts. Originally, the exogenous transcripts have been delivered to the cells by transfection with retroviral vectors, and the foreign DNA integrated into the cell own genome, which eventually silenced the viral DNA in a stable way. Since the first discovery, however, several other approaches have been elaborated and successfully employed to produce iPS cells reproducibly and efficiently from practically every cell type tried. Some of the technical issues have been overcome with the use of non-viral vectors as delivery tools [18], the use of non integrating DNA constructs [19], the reduction and variation on the number and identify of the transcripts needed [20], and the generation of iPS through small molecule induction and biochemical inhibitor cultures [21].

Because iPS can be derived from adult, fully differentiated cells, they offer two main advantages over ESC: first of all their derivation does not necessitate the destruction, or even the manipulation of preimplantation embryos. Since they can be derived from virtually any cell type, iPS cell lines with the same genotype of the cell donor can be derived, making it possible at least in theory to have custom and personalized cell lines for regenerative medicine.

## PLURIPOTENT CELLS IN DRUGS AND TOXICITY TESTING

The development of therapeutic drugs is a long process that starts from *in vitro* research on cellular models where chemicals with a potential beneficial effect for a pathological condition are identified; candidate molecules identified *in vitro* are first tested in animal models and finally the research is translated to clinical trials in humans; unfortunately, while humans and mice often respond similarly to chemicals, the interspecies variability makes animal models unreliable to a certain degree. To overcome this limitation, several tests for *in vitro* drug toxicity on human cells have been developed; before pluripotent stem cell derivation, a variety of somatic cell lines were used, mainly immortalized lines carrying chromosomal abnormalities and primary cultures. Unfortunately, the limitations of these tools for evaluating toxicity are many: first, primary cultures are characterized by a high degree of heterogeneity, generating inconsistent results; second, the immortalized lines deviate in their cellular response to stimuli from primary human cells, and therefore will respond to a chemical challenge in a different and possibly unrelated way; third, the response to certain agents can vary between different individuals from similar genetic background; moreover, to screen the early embryonic stages, no efficient human *in vitro* models are available. With the derivation of hESCs first and hiPSCs later, it became clear how pluripotent cells might represent an ideal source for drug development and toxicity testing. The capacity to maintain a proliferative state and differentiate in every cell type makes it possible, at least theoretically, to evaluate the effect of a certain molecule on a wide variety of tissues coming from the same genetic background; moreover, the possibility to reprogram adult cells into iPSCs and generate patient specific cell lines, allows to step forward into the "personalized medicine" field. Up to now many pluripotent-derived cells have been used to identify and test potential drugs although, mostly for a lack of consensus on the predictive value of these tests, none of the molecules identified has entered clinical trials yet. The bigger challenge for the widespread implantation of pluripotent cell-based *in vitro* drug discovery screenings is the relatively low efficiency of many differentiation protocols: to have accurate readouts, drug testing must be carried out on a homogeneous population of fully differentiated cells, and up to now there are few protocols that allow this kind of analysis [22].

## PLURIPOTENT CELL BASED DISEASE MODELS

One of the most relevant contributions of pluripotent stem cells to biomedical research so far is the possibility of modeling a disease *in vitro*. In several instances, the study of the pathogenesis of a disease has been historically challenging, since research has been limited by the scarcity of diseased biological material available from non invasive analysis. For instance, to study Alzheimer's disease development, one would have to rely on post-mortem samples or animal models, each presenting clear challenges. After the first report of the derivation of human induced pluripotent it was immediately clear that a new and exciting opportunity was developing: it was now possible to reprogram cells of an individual affected by a certain disease, differentiate them into the target cell type, and study the behavior of the cells. Soon, a wide array of iPS cell lines was derived from affected patients: Duchenne muscular dystrophy (DMD), Huntington disease (HD), diabetes mellitus type 1 (DM), Parkinson disease (PD) and Down syndrome (DS) [23]. Almost invariably, these cell lines showed the same genetic background of the patient, however, the reprogrammed cells did not always faithfully recapitulate the development of the disease, and while in some cases it has been possible to obtain predictive models, in others the cells lose their capacity to show the pathological phenotype. The most successful models are those referring to monogenic diseases showing a pathological phenotype, while a less stringent relationship was found for many multifactorial diseases [24].

Nonetheless, the technical possibility of having a "disease in a dish" opened the door for innovative studies in disease modeling such as Fanconi Anemia [13, 25], Gaucher's disease [26], Duchenne (DMD) and Becker muscular dystrophies (BMD) [23], with promising but still incomplete results.

While iPSCs allow the derivation of disease specific cells from selected patients, the situation is different for hESCs: to obtain disease specific ESC lines there is the need of a diseased embryo; for this reason the source for diseased hESC derivation are embryos positive after preimplantation genetic diagnosis (PGD). A drawback of this situation is the scarcity of material that limit the amount of diseases that can be studied; in this case only monogenic diseases can be taken into account: Huntington Disease [27], Charcot Marie Tooth type 1A, Cystic Fibrosis [28] and Fragile-X [29] are some examples of affected hESC lines derived in the last years. Most of the time the hESC lines derived had been shown to be genetically similar to diseased cells but few studies have been performed on the phenotype of the re-differentiated hESCs and not much is known about the efficacy of the models.

## PLURIPOTENT CELLS RESEARCH BY SYSTEMS AND ORGAN

### The Nervous System

Within the ectoderm-derived tissues the most challenging to study is without doubt the neural one; difficulties in obtaining samples hampered the research for long time, and developing human models to study neurodegenerative diseases has always been exceedingly difficult. Until recently, the only material available came from cadavers; it was therefore possible to evaluate only the endpoint of the different diseases but not much was known in term of pathophysiology and etiology. In 2009 it was reported the creation of an iPSC line from a patient affected by spinal muscular atrophy (SMA). The researchers were able to reprogram the patient's cells and then differentiate them into astrocytes and neurons; the cell line was lacking SMN1, the gene deleted in the pathology, and in the differentiated cells there was a selective death of motor neurons. Moreover, this same research served as a proof of principle for future drug development, because the authors were able to identify a molecule able to reduce the cell death by the *in vitro*, iPS derived neuronal assay [30].

Another work proved the successful reprogramming of adult cells from familial dysautonomia (FD); in this case the reprogrammed cells didn't show an impaired neural differentiation, but rather an impaired neurogenesis and a lower migration capacity, all features resembling the phenotype of the disease [25]. Again, in this case the disease modeled was monogenic with a clear phenotype. Parkinson disease (PD), on the other hand, is a complex multifactorial pathology in which genetic and environmental factors contribute to the loss of dopaminergic (DA) neurons in the brain [31]; several researchers derived iPSCs from PD patients and some of them exhibited the phenotype of the affected cells, paving the way to the development of cellular models for drug screening [32]. Unfortunately, each cell line showed only one specific subset of mutations characterizing specific subtypes of the pathology, hence only a minority of the molecular causes of PD can be screened at a time [33]. The same situation has been observed for other multifactorial diseases like Alzheimer's disease and amyotrophic lateral sclerosis (ALS) [34]; up to now there hasn't been any report of multifactorial diseases successfully modeled with pluripotent cells, since the combination of factors generating the pathology and their interactions is extremely complex; however, these models have been successfully used as a preliminary test for the development of new drugs targeting only specific forms of the pathology. One such example has been the identification of anacardic acid as candidate compound with therapeutical value for ALS [35]. In this study, iPSCs were derived from patients with a mutation in Tar DNA binding protein-43 (TDP-43). The iPSCs were then differentiated into motor neurons, which displayed the same phenotypic characteristics of the native diseased tissue; four compounds were tested on these cells, and the histone acetyltransferase inhibitor anacardic acid was able to rescue the phenotype of the diseased cells. PSC based *in vitro* models of ALS were also used to screen a library of small molecules to prevent motor neuron death; in this elegant work researchers tested the most promising hits from a mouse based screening on both iPSCs- and hESCs- derived motor neurons; the iPSCs were derived from patients carrying Super Oxide Dismutase 1 (SOD1) or TDP-43 mutations, while the hESCs were engineered to knock out the SOD1 gene. The effect of these compounds on motor neuron survival was compared in all the lines described; moreover, as control, the motor neuron survival was analyzed after treatment with molecules proven to be ineffective in ALS related clinical trials. The results obtained showed that the survival of the neurons was not improved after the treatment with ineffective drugs, while Kenpaullone, a GSK-3 inhibitor, increased significantly the survival of the motor neurons under different conditions [36]. These results are twofold striking; on the one hand they identify Kenpaullone as a potential candidate for the development of an effective treatment, while on the other hand show the discriminating ability of the *in vitro* model. These very promising results, anyway, are specific to the analyzed mutant cells, and ALS pathophysiology remains debated, with more findings needed to reach an agreement on the development of the disease.

## The Cardiovascular System

In the past decades several candidate drugs that entered clinical trials have been discontinued at a late stage of development because of cardiac toxicity. Heart diseases, especially those affecting the heart beating rate like arrhythmias, are poorly modeled with rodents because of the very different heart frequency, therefore *in vitro* testing on human cells is the preferred route [37]; cardiomyocytes are contracting cells, creating an additional problem for differentiation protocols, which must produce cardiomyocytes with the correct structural and electrophysiological characteristics. The protocols to obtain cardiomyocytes greatly improved in the last few years, resulting in the large-scale production and commercialization of the so called hSC-CMs (human stem cell derived cardiomyocytes) obtained from iPSCs and hESCs. In parallel, the development of new tools such as multi electrode array (MEA) for cardiomyocyte electrophysiology, high content analysis (HCA) to evaluate cell morphology, and impedance assays for beating/viability have allowed the assessment of cardio toxicity in hSC-CMs [38, 39]. In a recent study, 13 drugs with proven cardiac toxicity were tested on a hESCs-CMs population; the response of the cells after the treatment was analyzed combining the different structural and functional changes in the hESC-CMs and a multivariate analysis predicted the effect of the drugs with good accuracy, resembling in most of the cases the *in vivo* observation [40].

Cardiovascular diseases in general are within the most common causes of death in the developed world (WHO, Global status report on non communicable diseases 2014); for this reasons much effort has been put in the development of cellular models capable to mimic the development of the most common cardiac congenital diseases, since classical research in the cardiovascular field has the same drawback mentioned for the study of neural tissue, *i.e.* the scarcity of material available. In 2011 [41] it was shown that iPSCs can recapitulate the phenotype of long QT syndrome affected cells: when iPSCs were differentiated into cardiomyocytes they expressed the mutated form of KCNH2, one of the genes causing the pathology; furthermore, this cardiomyocytes showed a longer action potential when compared to a control population.

## Blood

Blood diseases are mostly being modeled with the aid of hiPSCs: many forms of leukemia, myeloproliferative neoplasms (MPNs), myelodysplastic syndrome, aplastic anemia, paroxysmal nocturnal hemoglobinuria can now be modeled thanks to the latest findings in term of cells reprogramming and the resulting iPSCs used for preliminary drug testing. Blood-derived iPSCs have been used, for example, to understand the lack of effectiveness of JAK inhibitors in the treatment of polycythemia vera (PV), a myeloproliferative neoplasm; in this study the researchers derived iPSCs from control individuals and from individuals affected by PV, then differentiated them into erythroblast and hematopoietic progenitors and observed the response of the differentiated progeny to target drugs already proven to be ineffective for the treatment of the disease. What they found is that the hematopoietic precursors from PV-iPSCs were not affected by the treatment, showing how this model would have correctly predicted the lack of effect of these drugs that were initially considered good candidates and were used in human trials [42].

## Liver, Pancreas and Lungs

Liver, pancreas, and pulmonary epithelium have been studied in depth to elucidate the mechanism of diseases like Type I [43] and Type II diabetes [44], Cystic fibrosis [45] and Familial Hypercholesterolemia (FH) [46]. In 2010 Rashid and colleagues derived patient specific iPSCs from patients suffering from Familial Hypercholesterolemia, alpha antitrypsin 1 deficiency and glycogen storage disease type 1A. From the three pathologies a further characterization revealed how the re-differentiated hepatocytes express a phenotype similar to the one of the diseased cells [46, 47]. Other modeled liver pathologies are Wilson disease [48], progressive familial hereditary cholestasis, tyrosinemia type 1 and Crigler-Najjar syndrome [46, 49]. The other main areas of interest are the diseases involving the respiratory tract; in cystic fibrosis, for example, it was found how iPSCs derived from F508del patients and differentiated into lung epithelial cells showed the incorrect localization of the CF found in the disease [50].

Because of the central role of the liver in determining the pharmacokinetics, bioavailability, and degradation of exogenous molecules, the process of drug discovery focus strongly on liver toxicity, the very reason behind the failure of one third of all candidate drugs under development. Hepatocytes have been derived from pluripotent cells since 2010 [51], and many studies now are using these cells as model to predict toxicity *in vitro*, from candidate drugs to any other kind of chemical; in 2011 the first report evaluated the effect of different chemical on a population of hESCs derived hepatocytes (hES-Hep) and analyzed the response of the cells through microarray analysis of gene expression [52]; the 15 compounds tested were representative of three categories: a) genotoxic carcinogens, b) non genotoxic carcinogens and c) non carcinogens. The hESCs derived hepatocytes were characterized and treated with the different chemicals; when analyzing the up or down regulation of specific pathways related to apoptosis and cell cycle the researchers were able to correctly classify the compounds in the three categories, showing how it is possible, despite just at an initial stage, to classify carcinogens and genotoxic agents with these tools; other studies used iPSCs derived hepatocytes and defined models of evaluation of chronic toxicity with promising results, showing how it is possible to evaluate also the effect of the compounds showing toxicity only after a long exposure time [53]. Unfortunately, the results are promising but preliminary and much more work is needed to obtain reliable models; up to now, in fact, protocols have not been standardized and the comparison of different studies using different sources of cells is very challenging.

## CELL THERAPY AND REGENERATIVE MEDICINE

The very same characteristics of pluripotent cells that makes them particularly interesting candidates for regenerative therapies are the ones that make their use in the clinic challenging; indefinite proliferation is a very useful characteristic when expanding cells *in vitro* but can become very dangerous if the cells are transplanted to a patient while retaining the characteristic of uncontrolled growth. Teratoma formation, a proof of pluripotency, is already indicative of the risks associated with the direct use of these cells, and what is needed is the generation, *in vitro*, of a fully differentiated progeny of mature cells or precursors, that only contributes to the regenerative process. As an alternative, pluripotent cells might be engineered to include fail-safe mechanisms which would activate cell death when certain genes of paramount importance for the undifferentiated state are expressed above a certain threshold.

Another important issue is the efficiency of the differentiation process: being the cells pluripotent, if not induced to differentiate in a very accurate and specific way, the resulting population will be a mix of cell types that will be of no practical use for further clinical application.

Generally speaking, the main factors to consider for tissue regeneration are:

1. The amount of cells needed to efficiently repair the organs.
2. The immunological compatibility of the graft.
3. The homogeneity of the population transplanted.
4. The "safety" of the transplanted cells (The absence of any kind of animal component in the process of derivation/differentiation, normal genetic background, safety of reprogramming procedures).
5. The route of delivery/transplantation.
6. Beneficial effect due to a proper tissue engraftment *vs.* increased regeneration of the original host tissue due to a paracrine effect of cytokines secreted by the transplanted cells.

All the above mentioned points are fundamental for the development of treatment options with pluripotent cells. With adult stem cells the concern about tolerance and tumorigenic potential are not particularly relevant since these cells can be derived from a patient and transplanted back to the same individual, avoiding the rejection, and the risk of developing tumors after a transplant is extremely low. Also iPSCs can be derived from the same patient that is going to receive the treatment, and in theory the rejection should be avoided; unfortunately, the process of reprogramming is long and expensive and this makes the "patient specific iPSCs" approach not realistic in a large scale setting. Furthermore, it is often unclear what the main cause of the beneficial effect observed in many studies is: most of the time, in fact, it seems that the cells transplanted are not able to engraft in the host tissue and after a short time they are not found in the target region anymore; this may imply that for the most part the beneficial effects observed were mediated by the cells secretome (anti-inflammatory molecules and cytokines) rather than a direct recovery of function based on an engraft-and-repair mechanism.

After many preclinical studies assessing the safety of these techniques in animal models, in 2010 the first human clinical trials using PSCs-derived cells to cure spinal cord injuries with hESC-derived oligodendrocytes received the green light from the FDA; in the same year two more trials were approved, to treat Age related Macular Degeneration (AMD) and Stargardt disease. Overall, clinical trials with pluripotent cells are not many, and most of them are still in phase I, *i.e.* testing the safety of the graft, so more time is needed to understand if these cells will become a reality in the clinic, anyway the preliminary data available gave promising results.

## Regenerative Therapies in the Eye

The eye is a privileged organ when speaking about models for transplantation: it is in fact able to tolerate foreign antigens without activating the immune response, it is easily accessible for transplantation and the progresses can be observed without the need of invasive techniques. As stated before, the first PSCs-derived cells to be transplanted were Retinal Pigmented Epithelium cells derived from hESCs, but many followed: a total of eight clinical trials is currently evaluating the effect of hESCs-derived RPEs in the cure of Wet and Dry AMD and Stargardt disease, using both cell suspension or cell sheets, while the Riken Institute in Japan is running the first clinical trial with iPSCs derived RPEs. The preliminary data of the study from Schwartz and colleagues [54] are very encouraging and report the absolute safety of the transplanted cells in all the patient treated; in addition midterm follow up studies showed an improved visual acuity (at different degrees) for 8 patients out of 18, stable acuity in 6 and a continuous visual acuity loss in 1. Another study [55] reports increased visual acuity for 3 out of 4 patients and a stable vision in 1.

# Heart Regeneration

The heart is an organ with a very limited regenerative capacity, and after myocardial infarction the cardiac cells are not able to quickly repopulate the infarcted area, which is instead replaced with scar tissue, eventually leading to heart failure. In 2015, the assessment of the safety and efficacy of cardiac precursors (CD15+ISL1+) derived from hESCs embedded in fibrin scaffold and transplanted into rat hearts has been reported; this progenitors resulted in an improved contractility and stabilized remodeling, ameliorating the heart function; these cells are now being investigated in a clinical trial actively recruiting patients with severe heart failure. Other previous studies [56] showed how hESC and iPSCs derived cardiac cells can contribute to the regeneration of the heart in pigs, the standard model for heart studies, although some problem of arrhythmias were observed in one of the studies, showing how the improvements are indeed remarkable but further studies are still needed.

# Hematopoietic Precursors

Many studies investigated the possibility of obtaining hematopoietic precursors from pluripotent cells to find a replacement for blood transfusion or bone marrow transplantation: despite the many progresses reported in the generation of erythrocytes, platelets, granulocyte or NK, the clinical application of PSCs in hematological diseases is still far, being the PSCs-derived blood cells generally immature, not functional and the differentiation protocols still inefficient [57]; one of the fundamental issues that needs to be addressed is the understanding of the interactions between the different molecular player that act during hematopoiesis, as highlighted by Sturgeon and colleagues, that showed how the modulation of Wnt can lead to the proper hematopoietic lineage specification during the differentiation of PSCs [58].

# Endodermal Tissues: Pancreas, Liver and Lungs

The regeneration of beta cells is one of the main goals in regenerative medicine, being Type I and Type II diabetes diseases affecting around 52 million people in Europe alone. PSCs differentiation has been thoroughly investigated in the last decade and up to now 2 groups reported encouraging results: the company

Viacyte reported the successful differentiation of hESCs into pancreatic progenitors [59, 60] and is currently evaluating the performance of these cells, named Pancreatic Endoderm Cells (PEC), in the cure of Type I Diabetes. Another group reported the derivation of mature beta cells from hESCs using a protocol suitable for large scale production and eventually applicable in the clinic [61].

Beside the pancreas, the liver is the most important organ of endodermal origin, especially studied to create *in vitro* models for toxicity screening (as already described); up to now a fully differentiated progeny of hepatocytes hasn't been obtained yet, but lately a combination of different populations of cells, namely hESCs supported by umbilical cord endothelial cells and mesenchymal stem cells, have proven to generate liver "buds", organoids able to rescue liver function in mice with liver failure [62]. Despite the studies being exclusively preclinical and in animal models, emerging evidences highlight the importance of this newly developed cells in the study of liver biology and toxicity screenings.

The study of lung diseases is hampered by the diversity in the structure and physiology of the organ between mice and humans; to date some protocols have been developed to generate alveolar epithelia type II from hESCs and iPSCs [63, 64], and lung epithelial cells have been proven to extend the lifespan and improve the pulmonary function of bleomycin-injured NOD/SCID immunodeficient mice that develop severe lung injury. Unfortunately for the clinic, the protocols still need to be optimized to achieve a more mature and homogeneous population, so up to now PSCs derived lung progenitors are mainly used in disease modeling studies.

**Central Nervous System**

Regenerative therapies are the only options for a real cure in neurodegenerative diseases, being pharmacological treatments only able to slow down or block the progression of the diseases and not useful to repopulate the degenerated organ; one clinical trial was authorized in 2010 for the treatment of patient that suffered from spinal cord injuries with hESCs derived oligodendrocytes; the trial was discontinued from GERON Corp., the company that initiated it, in 2011, but recently it was taken over by a different company, Asterias Biotherapeutics, that restarted it, therefore the results of the phase I still needs to be published. Apart from the one ongoing clinical trial many improvement have been made in the generation of neurons, especially dopaminergic neurons for the treatment of Parkinson Disease (PD). Recent works showed how DA neurons derived from both hESCs and iPSCs can integrate in the brain of 6-hydroxydopamine (6-OHDA) rats, an animal model for PD, and differentiate in DA-neurons able to form synapsis with specialized regions of the brain and capable of restoring the motor functions in the rodent model [65, 66]. Other neurodegenerative diseases have been studied and promising preclinical results were obtained, for example, in Alzheimer's Diesease (AD) treatment: in 2014 it was shown how macrophage-like myeloid cells differentiated from hiPSCs (iPSCs-ML), and engineered to express the Beta Amyloid peptide protease Neprilysin-2, are able to efficiently degrade toxic amyloid fibers in 5XFAD mice, standard model for AD characterized by 5 concomitant mutation [67]; this study also shows how alternative routes, beside the "classic" tissue engineering, can be pursued with PSCs. Another well studied disease is amyotrophic lateral sclerosis (ALS): in 2014 it was reported how hiPSCs differentiated into neurons could restore neuromuscular function and extend the lifespan of mouse models of ALS [68].

## CONCLUDING REMARKS

As pluripotent stem cell biology moves from the basic research laboratory to the clinic, we are witnessing more and more cell based applications being developed in such fields as drug discovery and testing, personalized toxicology, and clinical trials for cell based therapies in regenerative medicine for loss of function diseases. As we contemplate future stem cell based therapies to treat chronic and so far incurable diseases, we should pause and consider the importance of the development of human assisted reproduction on the very existence of pluripotent stem cells. *In vitro* preimplantation human embryology, and the development of cryopreservation techniques for human embryos, have allowed on one hand to fulfill the desire for a child in many infertile couples, and on the other for the donation of supernumerary embryos, not used in reproduction, to research. Without assisted reproduction there would be no embryonic stem cell derivation, and without the knowledge acquired by the study of hESC, the derivation of iPSc would have been almost impossible. The entire field of regenerative medicine, as a result, would have been of much reduced scope and promise.

## CONFLICT OF INTEREST

The authors confirm that the authors have no conflict of interest to declare for this publication.

## ACKNOWLEDGEMENTS

The authors wish to thank Sarai Brazal for technical editorial assistance.

## REFERENCES

[1]     Duncan SA, Nagy A, Chan W. Murine gastrulation requires HNF-4 regulated gene expression in the visceral endoderm: tetraploid rescue of Hnf-4(-/-) embryos. Development 1997; 124(2): 279-87.
        [PMID: 9053305]

[2]     Conley BJ, Young JC, Trounson AO, Mollard R. Derivation, propagation and differentiation of human embryonic stem cells. Int J Biochem Cell Biol 2004; 36(4): 555-67.
        [http://dx.doi.org/10.1016/j.biocel.2003.07.003] [PMID: 15010323]

[3]     Kleinsmith LJ, Pierce GB Jr. Multipotentiality of single embryonal carcinoma cells. Cancer Res 1964; 24(9): 1544-51.
        [PMID: 14234000]

[4]     Matsui Y, Zsebo K, Hogan BL. Derivation of pluripotential embryonic stem cells from murine primordial germ cells in culture. Cell 1992; 70(5): 841-7.
        [http://dx.doi.org/10.1016/0092-8674(92)90317-6] [PMID: 1381289]

[5]     Shamblott MJ, Axelman J, Wang S, *et al.* Derivation of pluripotent stem cells from cultured human primordial germ cells. Proc Natl Acad Sci USA 1998; 95(23): 13726-31.
        [http://dx.doi.org/10.1073/pnas.95.23.13726] [PMID: 9811868]

[6]     Thomson JA, Itskovitz-Eldor J, Shapiro SS, *et al.* Embryonic stem cell lines derived from human blastocysts. Science 1998; 282(5391): 1145-7.
        [http://dx.doi.org/10.1126/science.282.5391.1145] [PMID: 9804556]

[7]     Strelchenko N, Verlinsky O, Kukharenko V, Verlinsky Y. Morula-derived human embryonic stem cells. Reprod Biomed Online 2004; 9(6): 623-9.
        [http://dx.doi.org/10.1016/S1472-6483(10)61772-5] [PMID: 15670408]

[8]     Geens M, Mateizel I, Sermon K, *et al.* Human embryonic stem cell lines derived from single blastomeres of two 4-cell stage embryos. Hum Reprod 2009; 24(11): 2709-17.
        [http://dx.doi.org/10.1093/humrep/dep262] [PMID: 19633307]

[9]     Klimanskaya I, Chung Y, Becker S, Lu SJ, Lanza R. Human embryonic stem cell lines derived from single blastomeres. Nature 2006; 444(7118): 481-5.
        [http://dx.doi.org/10.1038/nature05142] [PMID: 16929302]

[10]    Vassena R, Montserrat N, Carrasco Canal B, *et al.* Accumulation of instability in serial differentiation

and reprogramming of parthenogenetic human cells. Hum Mol Genet 2012; 21(15): 3366-73.
[http://dx.doi.org/10.1093/hmg/dds168] [PMID: 22547223]

[11]   Lu B, Palacino J. A novel human embryonic stem cell-derived Huntingtons disease neuronal model exhibits mutant huntingtin (mHTT) aggregates and soluble mHTT-dependent neurodegeneration. FASEB J 2013; 27(5): 1820-9.
[http://dx.doi.org/10.1096/fj.12-219220] [PMID: 23325320]

[12]   Bosman A, Letourneau A, Sartiani L, *et al.* Perturbations of heart development and function in cardiomyocytes from human embryonic stem cells with trisomy 21. Stem Cells 2015; 33(5): 1434-46.
[http://dx.doi.org/10.1002/stem.1961] [PMID: 25645121]

[13]   Raya A, Rodríguez-Pizà I, Guenechea G, *et al.* Disease-corrected haematopoietic progenitors from Fanconi anaemia induced pluripotent stem cells. Nature 2009; 460(7251): 53-9.
[http://dx.doi.org/10.1038/nature08129] [PMID: 19483674]

[14]   Zhang N, An MC, Montoro D, Ellerby LM. Characterization of human Huntington's disease cell model from induced pluripotent stem cells. PLOS Currents Huntington Disease 2010.
[http://dx.doi.org/10.1371/currents.RRN1193]

[15]   Malan D, Friedrichs S, Fleischmann BK, Sasse P. Cardiomyocytes obtained from induced pluripotent stem cells with long-QT syndrome 3 recapitulate typical disease-specific features *in vitro.* Circ Res 2011; 109(8): 841-7.
[http://dx.doi.org/10.1161/CIRCRESAHA.111.243139] [PMID: 21799153]

[16]   Takahashi K, Yamanaka S. Induction of pluripotent stem cells from mouse embryonic and adult fibroblast cultures by defined factors. Cell 2006; 126(4): 663-76.
[http://dx.doi.org/10.1016/j.cell.2006.07.024] [PMID: 16904174]

[17]   Takahashi K, Tanabe K, Ohnuki M, *et al.* Induction of pluripotent stem cells from adult human fibroblasts by defined factors. Cell 2007; 131(5): 861-72.
[http://dx.doi.org/10.1016/j.cell.2007.11.019] [PMID: 18035408]

[18]   Gonzalez F, Barragan Monasterio M, Tiscornia G, *et al.* Generation of mouse-induced pluripotent stem cells by transient expression of a single nonviral polycistronic vector. Proc Natl Acad Sci USA 2009; 106(22): 8918-22.
[http://dx.doi.org/10.1073/pnas.0901471106] [PMID: 19458047]

[19]   Stadtfeld M, Nagaya M, Utikal J, Weir G, Hochedlinger K. Induced pluripotent stem cells generated without viral integration. Science 2008; 322(5903): 945-9.
[http://dx.doi.org/10.1126/science.1162494] [PMID: 18818365]

[20]   Giorgetti A, Montserrat N, Aasen T, *et al.* Generation of induced pluripotent stem cells from human cord blood using OCT4 and SOX2. Cell Stem Cell 2009; 5(4): 353-7.
[http://dx.doi.org/10.1016/j.stem.2009.09.008] [PMID: 19796614]

[21]   Shi Y, Desponts C, Do JT, Hahm HS, Schöler HR, Ding S. Induction of pluripotent stem cells from mouse embryonic fibroblasts by Oct4 and Klf4 with small-molecule compounds. Cell Stem Cell 2008; 3(5): 568-74.
[http://dx.doi.org/10.1016/j.stem.2008.10.004] [PMID: 18983970]

[22]   Anson BD, Kolaja KL, Kamp TJ. Opportunities for use of human iPS cells in predictive toxicology.

Clin Pharmacol Ther 2011; 89(5): 754-8.
[http://dx.doi.org/10.1038/clpt.2011.9] [PMID: 21430658]

[23] Park IH, Arora N, Huo H, *et al.* Disease-specific induced pluripotent stem cells. Cell 2008; 134(5): 877-86.
[http://dx.doi.org/10.1016/j.cell.2008.07.041] [PMID: 18691744]

[24] Siller R, Greenhough S, Park IH, Sullivan GJ. Modelling human disease with pluripotent stem cells. Curr Gene Ther 2013; 13(2): 99-110.
[http://dx.doi.org/10.2174/1566523211313020004] [PMID: 23444871]

[25] Lee G, Papapetrou EP, Kim H, *et al.* Modelling pathogenesis and treatment of familial dysautonomia using patient-specific iPSCs. Nature 2009; 461(7262): 402-6.
[http://dx.doi.org/10.1038/nature08320] [PMID: 19693009]

[26] Panicker LM, Miller D, Park TS, *et al.* Induced pluripotent stem cell model recapitulates pathologic hallmarks of Gaucher disease. Proc Natl Acad Sci USA 2012; 109(44): 18054-9.
[http://dx.doi.org/10.1073/pnas.1207889109] [PMID: 23071332]

[27] Bradley CK, Scott HA, Chami O, *et al.* Derivation of Huntingtons disease-affected human embryonic stem cell lines. Stem Cells Dev 2011; 20(3): 495-502.
[http://dx.doi.org/10.1089/scd.2010.0120] [PMID: 20649476]

[28] Pickering SJ, Minger SL, Patel M, *et al.* Generation of a human embryonic stem cell line encoding the cystic fibrosis mutation deltaF508, using preimplantation genetic diagnosis. Reprod Biomed Online 2005; 10(3): 390-7.
[http://dx.doi.org/10.1016/S1472-6483(10)61801-9] [PMID: 15820050]

[29] Urbach A, Bar-Nur O, Daley GQ, Benvenisty N. Differential modeling of fragile X syndrome by human embryonic stem cells and induced pluripotent stem cells. Cell Stem Cell 2010; 6(5): 407-11.
[http://dx.doi.org/10.1016/j.stem.2010.04.005] [PMID: 20452313]

[30] Ebert AD, Yu J, Rose FF Jr, *et al.* Induced pluripotent stem cells from a spinal muscular atrophy patient. Nature 2009; 457(7227): 277-80.
[http://dx.doi.org/10.1038/nature07677] [PMID: 19098894]

[31] Damier P, Hirsch EC, Agid Y, Graybiel AM. The substantia nigra of the human brain. II. Patterns of loss of dopamine-containing neurons in Parkinsons disease. Brain 1999; 122(Pt 8): 1437-48.
[http://dx.doi.org/10.1093/brain/122.8.1437] [PMID: 10430830]

[32] Badger JL, Cordero-Llana O, Hartfield EM, Wade-Martins R. Parkinsons disease in a dish - Using stem cells as a molecular tool. Neuropharmacology 2014; 76(Pt A): 88-96.
[http://dx.doi.org/10.1016/j.neuropharm.2013.08.035] [PMID: 24035919]

[33] Cooper O, Seo H, Andrabi S, *et al.* Pharmacological rescue of mitochondrial deficits in iPSC-derived neural cells from patients with familial Parkinson's disease. Sci Trans Med 2012; 4(141): 141ra90.
[http://dx.doi.org/10.1126/scitranslmed.3003985]

[34] Wan W, Cao L, Kalionis B, Xia S, Tai X. Applications of induced pluripotent stem cells in studying the neurodegenerative diseases. Stem Cells Int 2015; 382530.

[35] Egawa N, Kitaoka S, Tsukita K, *et al.* Drug screening for ALS using patient-specific induced pluripotent stem cells. Sci Trans Med 2012; 4(145): 145ra104.

[http://dx.doi.org/10.1126/scitranslmed.3004052]

[36]  Yang YM, Gupta SK, Kim KJ, *et al.* A small molecule screen in stem-cell-derived motor neurons identifies a kinase inhibitor as a candidate therapeutic for ALS. Cell Stem Cell 2013; 12(6): 713-26. [http://dx.doi.org/10.1016/j.stem.2013.04.003] [PMID: 23602540]

[37]  Hoekstra M, Mummery CL, Wilde AA, Bezzina CR, Verkerk AO. Induced pluripotent stem cell derived cardiomyocytes as models for cardiac arrhythmias. Front Physiol 2012; 3: 346. [http://dx.doi.org/10.3389/fphys.2012.00346] [PMID: 23015789]

[38]  Pointon A, Abi-Gerges N, Cross MJ, Sidaway JE. Phenotypic profiling of structural cardiotoxins *in vitro* reveals dependency on multiple mechanisms of toxicity. Toxicol Sci 2013; 132(2): 317-26. [http://dx.doi.org/10.1093/toxsci/kft005] [PMID: 23315586]

[39]  Clements M, Thomas N. High-throughput multi-parameter profiling of electrophysiological drug effects in human embryonic stem cell derived cardiomyocytes using multi-electrode arrays. Toxicol Sci 2014; 140(2): 445-61. [http://dx.doi.org/10.1093/toxsci/kfu084] [PMID: 24812011]

[40]  Clements M, Millar V, Williams AS, Kalinka S. Bridging functional and structural cardiotoxicity assays using human embryonic stem-cell derived cardiomyocytes for a more comprehensive risk assessment. Toxicol Sci 2015; 148(1): 241-60. [http://dx.doi.org/10.1093/toxsci/kfv180] [PMID: 26259608]

[41]  Itzhaki I, Maizels L, Huber I, *et al.* Modelling the long QT syndrome with induced pluripotent stem cells. Nature 2011; 471(7337): 225-9. [http://dx.doi.org/10.1038/nature09747] [PMID: 21240260]

[42]  Ye Z, Liu CF, Lanikova L, *et al.* Differential sensitivity to JAK inhibitory drugs by isogenic human erythroblasts and hematopoietic progenitors generated from patient-specific induced pluripotent stem cells. Stem Cells 2014; 32(1): 269-78. [http://dx.doi.org/10.1002/stem.1545] [PMID: 24105986]

[43]  Maehr R, Chen S, Snitow M, *et al.* Generation of pluripotent stem cells from patients with type 1 diabetes. Proc Natl Acad Sci USA 2009; 106(37): 15768-73. [http://dx.doi.org/10.1073/pnas.0906894106] [PMID: 19720998]

[44]  Ohmine S, Squillace KA, Hartjes KA, *et al.* Reprogrammed keratinocytes from elderly type 2 diabetes patients suppress senescence genes to acquire induced pluripotency. Aging (Albany, NY) 2012; 4(1): 60-73. [http://dx.doi.org/10.18632/aging.100428] [PMID: 22308265]

[45]  Mou H, Zhao R, Sherwood R, *et al.* Generation of multipotent lung and airway progenitors from mouse ESCs and patient-specific cystic fibrosis iPSCs. Cell Stem Cell 2012; 10(4): 385-97. [http://dx.doi.org/10.1016/j.stem.2012.01.018] [PMID: 22482504]

[46]  Rashid ST, Corbineau S, Hannan N, *et al.* Modeling inherited metabolic disorders of the liver using human induced pluripotent stem cells. J Clin Invest 2010; 120(9): 3127-36. [http://dx.doi.org/10.1172/JCI43122] [PMID: 20739751]

[47]  Cayo MA, Cai J, DeLaForest A, *et al.* JD induced pluripotent stem cell-derived hepatocytes faithfully recapitulate the pathophysiology of familial hypercholesterolemia. Hepatology 2012; 56(6): 2163-71.

[http://dx.doi.org/10.1002/hep.25871] [PMID: 22653811]

[48] Yi F, Qu J, Li M, *et al.* Establishment of hepatic and neural differentiation platforms of Wilsons disease specific induced pluripotent stem cells. Protein Cell 2012; 3(11): 855-63.
[http://dx.doi.org/10.1007/s13238-012-2064-z] [PMID: 22806248]

[49] Ghodsizadeh A, Taei A, Totonchi M, *et al.* Generation of liver disease-specific induced pluripotent stem cells along with efficient differentiation to functional hepatocyte-like cells. Stem Cell Rev 2010; 6(4): 622-32.
[http://dx.doi.org/10.1007/s12015-010-9189-3] [PMID: 20821352]

[50] Wong AP, Bear CE, Chin S, *et al.* Directed differentiation of human pluripotent stem cells into mature airway epithelia expressing functional CFTR protein. Nat Biotechnol 2012; 30(9): 876-82.
[http://dx.doi.org/10.1038/nbt.2328] [PMID: 22922672]

[51] Brolén G, Sivertsson L, Björquist P, *et al.* Hepatocyte-like cells derived from human embryonic stem cells specifically *via* definitive endoderm and a progenitor stage. J Biotechnol 2010; 145(3): 284-94.
[http://dx.doi.org/10.1016/j.jbiotec.2009.11.007] [PMID: 19932139]

[52] Yildirimman R, Brolén G, Vilardell M, *et al.* Human embryonic stem cell derived hepatocyte-like cells as a tool for *in vitro* hazard assessment of chemical carcinogenicity. Toxicol Sci 2011; 124(2): 278-90.
[http://dx.doi.org/10.1093/toxsci/kfr225] [PMID: 21873647]

[53] Holmgren G, Sjögren AK, Barragan I, *et al.* Long-term chronic toxicity testing using human pluripotent stem cell-derived hepatocytes. Drug Metab Dispos 2014; 42(9): 1401-6.
[http://dx.doi.org/10.1124/dmd.114.059154] [PMID: 24980256]

[54] Schwartz SD, Regillo CD, Lam BL, *et al.* Human embryonic stem cell-derived retinal pigment epithelium in patients with age-related macular degeneration and Stargardt's macular dystrophy: follow-up of two open-label phase 1/2 studies. Lancet 2015; 385(9967): 509-16.
[http://dx.doi.org/10.1016/S0140-6736(14)61376-3] [PMID: 25458728]

[55] Song WK, Park KM, Kim HJ, *et al.* Treatment of macular degeneration using embryonic stem cell-derived retinal pigment epithelium: preliminary results in Asian patients. Stem Cell Rep 2015; 4(5): 860-72.
[http://dx.doi.org/10.1016/j.stemcr.2015.04.005] [PMID: 25937371]

[56] Ye L, Chang YH, Xiong Q, *et al.* Cardiac repair in a porcine model of acute myocardial infarction with human induced pluripotent stem cell-derived cardiovascular cells. Cell Stem Cell 2014; 15(6): 750-61.
[http://dx.doi.org/10.1016/j.stem.2014.11.009] [PMID: 25479750]

[57] Ackermann M, Liebhaber S, Klusmann JH, Lachmann N. Lost in translation: pluripotent stem cell-derived hematopoiesis. EMBO Mol Med 2015; 7(11): 1388-402.
[http://dx.doi.org/10.15252/emmm.201505301] [PMID: 26174486]

[58] Sturgeon CM, Ditadi A, Awong G, Kennedy M, Keller G. Wnt signaling controls the specification of definitive and primitive hematopoiesis from human pluripotent stem cells. Nat Biotechnol 2014; 32(6): 554-61.
[http://dx.doi.org/10.1038/nbt.2915] [PMID: 24837661]

[59] DAmour KA, Bang AG, Eliazer S, *et al.* Production of pancreatic hormone-expressing endocrine cells

from human embryonic stem cells. Nat Biotechnol 2006; 24(11): 1392-401.
[http://dx.doi.org/10.1038/nbt1259] [PMID: 17053790]

[60]   Kroon E, Martinson LA, Kadoya K, *et al.* Pancreatic endoderm derived from human embryonic stem
       cells generates glucose-responsive insulin-secreting cells *in vivo.* Nat Biotechnol 2008; 26(4): 443-52.
       [http://dx.doi.org/10.1038/nbt1393] [PMID: 18288110]

[61]   Pagliuca FW, Millman JR, Gürtler M, *et al.* Generation of functional human pancreatic β cells *in vitro.*
       Cell 2014; 159(2): 428-39.
       [http://dx.doi.org/10.1016/j.cell.2014.09.040] [PMID: 25303535]

[62]   Takebe T, Sekine K, Enomura M, *et al.* Vascularized and functional human liver from an iPSC-
       derived organ bud transplant. Nature 2013; 499(7459): 481-4.
       [http://dx.doi.org/10.1038/nature12271] [PMID: 23823721]

[63]   Wang D, Haviland DL, Burns AR, Zsigmond E, Wetsel RA. A pure population of lung alveolar
       epithelial type II cells derived from human embryonic stem cells. Proc Natl Acad Sci USA 2007;
       104(11): 4449-54.
       [http://dx.doi.org/10.1073/pnas.0700052104] [PMID: 17360544]

[64]   Ghaedi M, Calle EA, Mendez JJ, *et al.* Human iPS cell-derived alveolar epithelium repopulates lung
       extracellular matrix. J Clin Invest 2013; 123(11): 4950-62.
       [http://dx.doi.org/10.1172/JCI68793] [PMID: 24135142]

[65]   Doi D, Samata B, Katsukawa M, *et al.* Isolation of human induced pluripotent stem cell-derived
       dopaminergic progenitors by cell sorting for successful transplantation. Stem Cell Rep 2014; 2(3):
       337-50.
       [http://dx.doi.org/10.1016/j.stemcr.2014.01.013] [PMID: 24672756]

[66]   Grealish S, Diguet E, Kirkeby A, *et al.* Human ESC-derived dopamine neurons show similar
       preclinical efficacy and potency to fetal neurons when grafted in a rat model of Parkinsons disease.
       Cell Stem Cell 2014; 15(5): 653-65.
       [http://dx.doi.org/10.1016/j.stem.2014.09.017] [PMID: 25517469]

[67]   Takamatsu K, Ikeda T, Haruta M, *et al.* Degradation of amyloid beta by human induced pluripotent
       stem cell-derived macrophages expressing Neprilysin-2. Stem Cell Res (Amst) 2014; 13(3 Pt A): 442-
       53.
       [http://dx.doi.org/10.1016/j.scr.2014.10.001] [PMID: 25460605]

[68]   Nizzardo M, Simone C, Rizzo F, *et al.* Minimally invasive transplantation of iPSC-derived
       ALDHhiSSCloVLA4+ neural stem cells effectively improves the phenotype of an amyotrophic lateral
       sclerosis model. Hum Mol Genet 2014; 23(2): 342-54.
       [http://dx.doi.org/10.1093/hmg/ddt425] [PMID: 24006477]

# CHAPTER 5

# Exploiting the Role of Hematopoietic Stem Cell Transplantation as a Cure of Hematological and Non Hematological Diseases

Katia Perruccio*, Elena Mastrodicasa, Francesco Arcioni, Ilaria Capolsini, Carla Cerri, Grazia Maria Immacolata Gurdo and Maurizio Caniglia

*Pediatric Oncology Hematology Unit, Perugia General Hospital, Località Sant'Andrea delle Fratte, 06156 Perugia, Italy*

**Abstract:** This chapter describes hematopoietic stem cells and their therapeutic uses to cure otherwise lethal, malignant and non-malignant diseases. Here we analyze the biological characteristics of different hematopoietic stem cell sources and how they are mobilized, collected, selected from the patient himself for autologous transplantation, or from matched or mismatched, related or unrelated donors for allogeneic transplantation [1 - 3]. Hematopoietic stem cell transplantation has been implemented for decades and has undergone many improvements over the years [3]. Today it is a safe, feasible option for selected patients and it still remains the only cure for a wide range of malignancies or non-malignant diseases despite advances in understanding disease genetics and biology. Moreover, with improvements in conditioning regimens and graft manipulation [2, 4], cells can be transplanted to enhance immune reconstitution and reduce relapse, which are the most common cause of transplant failure [2, 4]. Given the immunological modulation and anti-leukemic effect of these cells, conditioning regimen can be reduced and transplantation is now extended to elderly patients who are more susceptible to drug toxicity.

**Keywords:** Autoimmune diseases, Hematological diseases, Immunology, Solid tumors, Stem cell transplantation, Stem cells.

---

* **Corresponding author Katia Perruccio:** Pediatric Oncology Hematology Unit, Perugia General Hospital, Località Sant'Andrea delle Fratte, 06156 Perugia, Italy; Tel: +39 075 5782415-2202; Fax: +39 075 5782204; E-mail: katia.perruccio@ospedale.perugia.it

Nicola Daniele (Ed.)

## INTRODUCTION

The discovery of several types of stem cells (SCs) that have different roles in human tissue creation and regeneration, makes them potentially promising as treatment for degenerative diseases such as myocarditis [5], Alzheimer's and Parkinson's disease, and other degenerative neurological disorders and autoimmune diseases [6], for which currently available therapies are ineffective in the long term. *In vitro* and *in vivo* human SCs have demonstrated great capacities for tissue regeneration and immune system modulation [7]. The regenerative capacity of hematopoietic stem cells (HSCs) from bone marrow (BM), peripheral blood (PB) and umbilical cord blood (UCB) [8, 9] has been known for decades. Over time technological advances in harvesting SCs, separating sub-populations with specific functions, and conservation have become more and more specialized and automated, so that today, worldwide, SC therapy is safe and successful for a wide range of human diseases like, for example, acute leukemia (AL), multiple myeloma (MM), severe aplastic anemia (SAA), thalassemia major and other hemoglobin diseases, immune-deficiencies and some solid tumors. This chapter will focus on HSC properties and their therapeutic uses, allogeneic (from a donor) and autologous (self) transplantation and future perspectives.

### Hematopoietic Stem Cells

Hematopoiesis initially occurs in long and flat bones, and then in vertebrae, the sternum, ribs and iliac crests [10]. Based on an average adult blood volume of about 5 liters, each day an adult human produces $2 \times 10^{11}$ erythrocytes, $1 \times 10^{11}$ leukocytes and $1 \times 10^{11}$ platelets which all derive from HSCs. Primitive, multi-potent HSCs are rare, occurring at a frequency of 1 in 10,000 to 1,000,000 BM cells, are quiescent in the steady-state, and at any one time, only a fraction enter the cell cycle to proliferate and give rise to different progenitors. They gradually lose one or more developmental potentials and ultimately become committed to a single cell lineage, which matures into the corresponding blood cell type. With the advent of monoclonal antibodies and flow cytometry, HSCs and their progenitors are characterized by the presence or absence of specific surface markers and by the ability to efflux fluorescent dyes [11]. The typical HSC marker is CD34, an integral trans-membrane glycoprotein with a molecular weight of 105-120 Kd (354 amino acids) which is expressed in 1-3% of BM cells, in 0.01-0.1% of PB cells, and in 0.1-0.4% of UCB cells [12]. Identifying BM stem and progenitor cells, CD34 is a maturation, but not a differentiation marker. Neither a receptor nor a signal transduction molecule, it is involved in SC homing and adhesion processes [13]. Synthesis of monoclonal antibodies against the CD34 antigen has provided the means for developing immune-magnetic systems to separate HSCs from blood. Given the paucity of CD34+ HSCs, hematopoietic growth-factors like Granulocyte-Colony Stimulating Factor (G-CSF) are administered to patients before autologous SCT (ASCT), and to donors before allogeneic SCT [14]. In the BM, hematopoietic growth-factors expand the SC compartment and alter adhesion to the stroma. CD34+ hematopoietic progenitors increase in number to the point that they leave the BM microenvironment and pour into the bloodstream (mobilization), where they are easily collected by leukapheresis. Selection procedures separate ontogenetically more immature progenitors, characterized by self-renewal and multi-lineage differentiation capacity, which are able to reconstitute the recipient's hematopoietic system [14]. Reconstitution depends on HSC "homing" as they migrate in the recipient from PB to the BM microenvironment. SC homing is a multi-step process: 1) Rolling on vascular endothelium. PBSCs (for example after a graft infusion) weakly bind adhesion molecules (selectins E and P) which are constitutively expressed by the vascular endothelium. These weak links cyclically create and break, allowing cells to roll on endothelium surface [15]. 2) Activation. After interaction with vascular endothelium, SCs that do not express the CXCR4 receptor detach from endothelium and return into circulation, while those expressing the CXCR4 receptor are activated by CXCL12 and SDF1, cytokines produced by BM endothelial cells. Links with adhesion molecules LFA-1/ICAM-1 and VLA-4/VCAM-1 [16] are stimulated to block SCs on the endothelial surface. 3) Migration from blood vessel to tissue. VLA-4 and VLA-5 integrins interact with fibronectin in the extracellular matrix, facilitating CXCR4+ stem cell migration through endothelial barrier fenestrations and penetration across the underlying basal lamina. 4) Homing to the BM microenvironment. SCs finally reach the BM niche. This microenvironment, which maintains SC properties and quiescence, is populated by various types of cells, adhesion molecules, chemotactic and growth factors [17]. This complex and fascinating process is schematically illustrated in Fig. (**1**).

## Bone Marrow Microenvironment

The BM niche is where HSCs home and self-renew and where they are protected from differentiation. In the BM niche, the SC number is strictly regulated and their maintenance depends on interactions with other cells, activating or inhibiting signal transductions, all of which are responsible for SC expansion [18]. All cells which reach the niche can assume SC potential, even for example, cancer cells. Malignant SCs are quiescent, excrete toxic drugs by ATP-binding transporters, repair DNA efficiently and resist apoptosis [19, 20]. Eradication of these rare leukemic SCs is essential if leukemia is to be cured [19, 20]. Fig. (**2**) illustrates a BM niche and hemopoiesis.

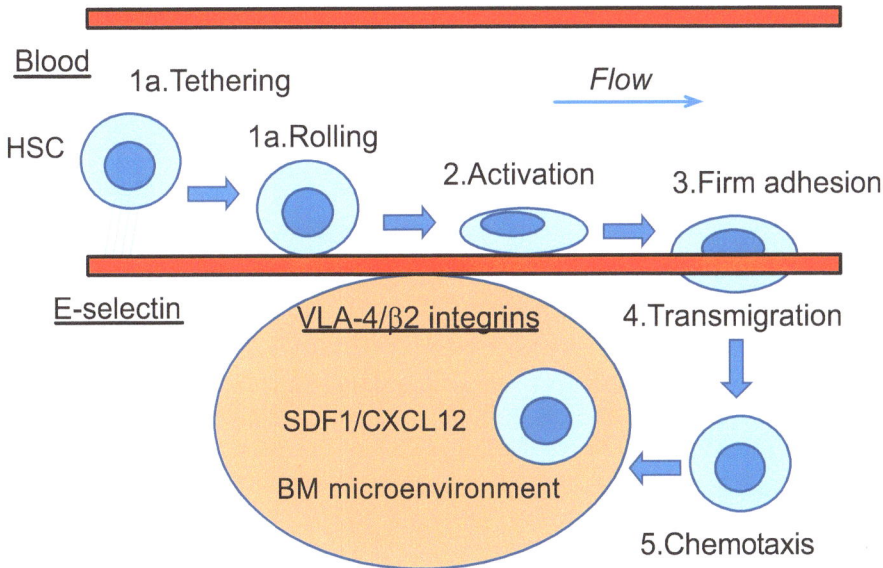

**Fig. (1).** Schema of HSC migration, showing the central roles of endothelial E-selectin and of SDF-1/CXCL12, chemokines that mediate HSC transmigration and seeding within BM microenvironment.

The BM niche not only provides a support for HSCs, but also regulates the behavior of cells that occupy it [21, 22]. Once identified as BM stroma, cells which constitute the bone tissue are now known to be adipocytes, osteoblasts (which play a key role in HSC regulation), macrophages, mesenchymal, endothelial and muscle cells. All these cells contribute to build a unique supporting architecture, which secretes cytokines and chemokines that are fundamental in the hematopoietic process. This is controlled by cell line specific growth factors and their receptors, as well as transcriptional factors regulating myeloid or lymphoid line-specific genes [22]. The tyrosine-kinase (TK) receptor class plays a major role in hemopoiesis. More than 30 TK receptors are expressed on the CD34+ cell surface, like for example platelet derived growth factor receptor (PDGFR) [23], monocyte-colony stimulating factor (M -CSF) receptor [24], Kit ligand (KL), stem cell factor (SCF) receptor and FLT3 receptors [25].

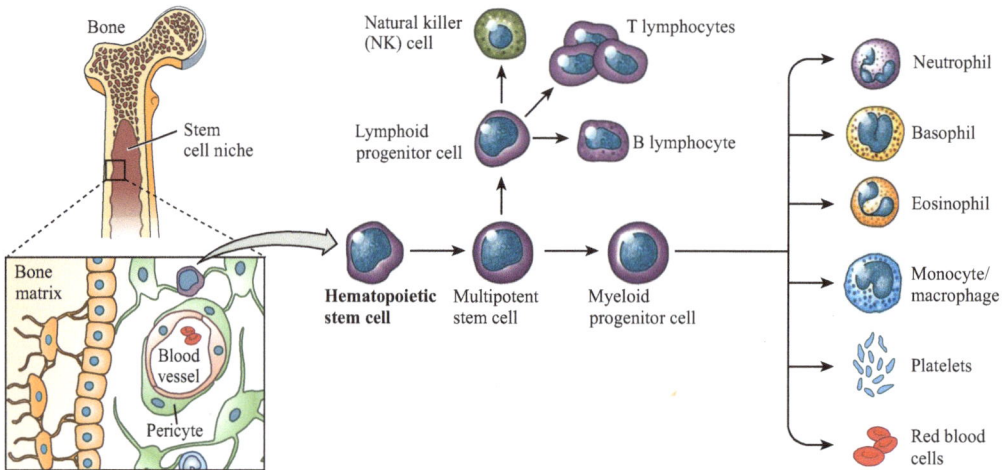

**Fig. (2).** Hematopoietic stem cells in the BM niche. The HSC is located in the bone marrow and generates a second SC that becomes either a lymphocyte progenitor cell giving rise to T, B, and Natural Killer (NK) cells, or a myeloid progenitor cell, which generates granulocytes, red blood cells and platelets [21].

These TK receptors link to a specific extracellular substrate to induce intracellular signals through phosphorylation of intermediate proteins. On the other hand, receptors without enzymatic activity, which are generally comprised of hetero-dimers with an alpha and a beta subunit, also play a role in hematopoiesis. The alpha subunit links the ligand, and the beta subunit conveys the intra-cellular signal through interactions with intra-cellular kinases like Src or JAK. This activation cascade modifies cell gene expression and induces progenitor differentiation and survival, depending on which transcription factors are involved [26]. The first transcription factor to be identified in this cascade was GATA-1

[27] which induces progenitor cells to differentiate into erythroblasts and megakaryocytes, by inhibiting the PU.1 transcription factor. Involved in HSC differentiation into myeloid precursors, PU.1 activation leads to monocyte and macrophage production. If, on the other hand, PU.1 is inhibited by C/EBP, another transcription factor, SCs differentiate into granulocytes [28]. Fig. (**3**) summarizes this process.

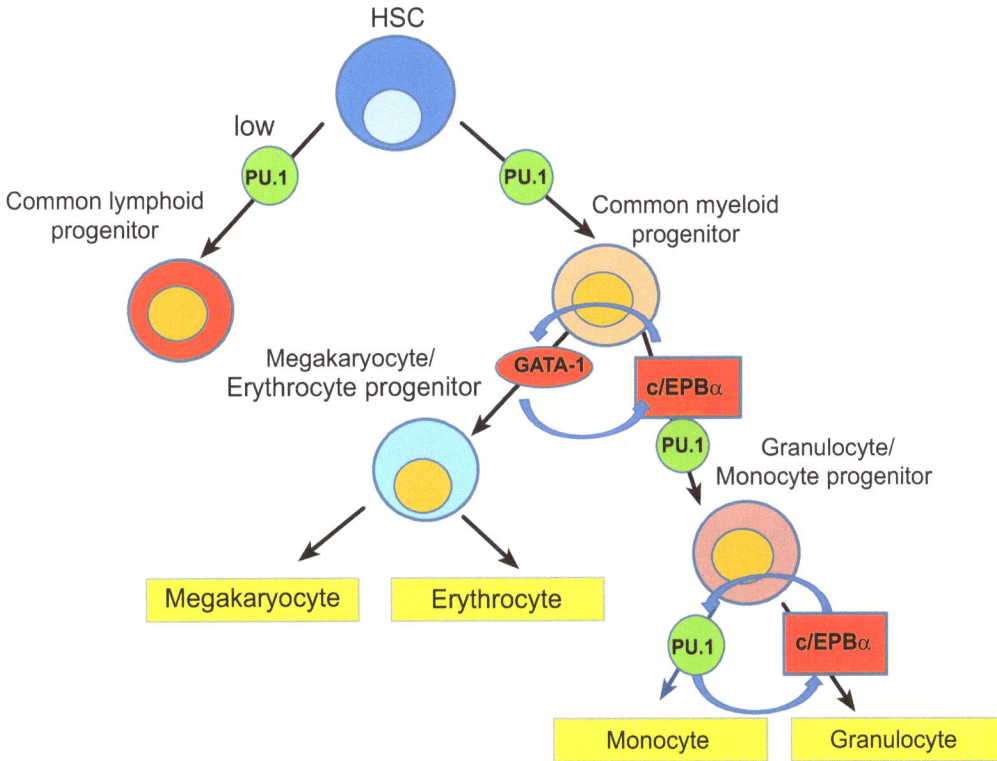

**Fig. (3).** Schema of GATA-1, PU.1 and C/EBPα involvement in the hematopoietic system differentiation.

## Hematopoietic Stem Cell Transplantation

As described above, CD34+ cell properties (migration, collection, homing) opened new perspectives in the use of BMT for the treatment of hematological malignancies. In 1939 a woman with gold-induced aplasia was given BM intravenously from her brother with the same blood group. This first human BMT transplant was not successful, and the patient died five days later [29]. After many other attempts, BM was successfully transplanted for the first time in 1968 in a child with Wiskott-Aldrick syndrome [30 - 32]. BMT success is due to high-dose chemotherapy and irradiation in the conditioning regimen [33] which has a fourfold purpose: to ensure leukemia de-bulking and overcome drug resistance, to prevent rejection by weakening/suppressing the patient's immune system, to allow the graft-*versus*-tumor (GvT) effect which is donor immune mediated eradication of leukemia. In the allogeneic setting, the GvT effect is central to BMT success [33]. Between 1969 and 1975, more and more patients with AL or SAA received BMT from identical twin donors [34] or histocompatibility leukocyte antigen (HLA)-matched siblings with many becoming long-term, disease-free survivors [35]. Outcomes after allogeneic matched sibling BMT, with the only barriers being "minor" HLA antigens [36], and syngeneic identical twin BMT, showed the relapse rate was significantly higher after syngeneic BMT [37] because lack of minor HLA disparities precluded the GvT effect. Alloreactive donor T cells in the graft [38, 39] are however a two-edged sword. On the one hand they exert the GvT or graft-*versus*-leukemia (GvL) effect, but on the other they trigger Graft-*versus*-host disease (GvHD) [4, 39, 40] an entity which has emerged only with transplant procedures. Donor T- cells recognize HLA antigens not only on leukemic cells but also on normal tissues and thus they damage organs like the skin, liver, bowel and thymus, consequently worsening post-transplant immune reconstitution [41, 42]. Since GvHD is lethal if left untreated, HSCT from un-manipulated matched grafts requires pharmacological immune suppression to prevent GvHD. Commonly used agents like Cyclosporine A (CSA), short-term Methotrexate (MTX), and so on delay immune recovery and increase infectious morbidity and mortality. Over the last 40 years, HSCT from HLA matched siblings has become treatment of choice for many hematological diseases [35 - 37]. With modern techniques for immune-suppression, transplants usually result in rapid engraftment, a relatively low risk of fatal GvHD, and successful development of immune tolerance and immune reconstitution [43]. Lack of a matched sibling as donor is the main limitation.

Any given sibling pair has only a 25% chance of inheriting the same HLA haplotypes from their parents and as there are, on average, fewer than three siblings per family in most developed countries and one is the patient, under 40% of patients will have a matched sibling [8, 44]. Consequently, world-wide volunteer donor registries were established for HLA-matched, unrelated transplants (MUD) with the aim of providing an unrelated donor with a similar HLA phenotype to the patient's [45]. All the volunteer donor registries regulate HLA-typing, donor selection, coordinate donor searches, BM harvest and transplantation. They form an international network, the Bone Marrow Donor Worldwide (BMDW), so that the donor search can be extended all over the world. Recent multivariate analyses of transplant outcomes in 2,223 adults with acute myeloid leukemia (AML) showed similar survival times for matched related or unrelated HSCT [46]. At the end of 1980's, UCB was proposed as another source of HCSs. The first UCB transplant was performed in Paris for a child with Fanconi' s Anemia [47]. Subsequently, UCB was extended to other malignancies but at present, is limited to the pediatric population because of the low CD34+ cell content in each cord blood unit which is ten-fold lower than in other HSC sources. The advantages of UCB include long-term storage in appropriate banks, immediate availability, and lymphocyte naivety which is associated with a lower risk of GvHD as it allows an HLA-disparity of up to 2 alleles [4-5/6] on HLA-A, B and DRB1 loci [48]. Furthermore, UCB stem cells provide a much greater BM repopulation capacity than stem cells from other sources and a lower transmission risk of donor latent infections, like Cytomegalovirus (CMV) and Epstein-Barr Virus (EBV) [49].

Due to the low CD34+ cell content, UCBT was associated with 10-25% rate of graft failure. Proposals to overcome this drawback include infusion of two UCB units, reduced intensity conditioning (RIC) regimens, or *in vitro* expansion of UCB-SCs [50 - 52]. Despite over 7 million HLA-typed volunteers in the donor registries and UCB banks worldwide, the chance of finding a MUD depends on HLA diversity and varies with race, ranging from 60-70% in Caucasians to under 10% for ethnic minorities [4, 8, 35, 36, 44]. Other limitations are the time lapse from registering to identifying a MUD (usually three months) which may lead to disease progression in patients who urgently need a transplant *e.g.* those with acute leukemia [4, 44, 45].

Paradoxically, closer DNA matching restricts the pool of donors, thus reducing the chances of finding a suitable MUD [53]. For all these reasons MUD HSCT is not feasible for over half of the patients requiring transplantation and so attention has focused on family members other than the HLA-matched sibling. Almost all patients have at least one HLA- haploidentical mismatched family member (parent, child, sibling or cousin and so forth), who is immediately available as donor. Unfortunately, full haplotype mismatched transplants were associated with an extremely high incidence of severe acute GvHD [54]. Although GvHD was prevented by extensive T-cell depletion [55, 56] the rejection rates rose steeply [57 - 60] because the balance between competing recipient and donor T cells shifted in favor of the unopposed host-*versus*-graft reaction. The barrier to engraftment of T-cell-depleted mismatched transplants was first overcome in 1993 with the clinical application of the SC mega-dose, a principle successfully pioneered in experimental models in the late 1980's [61 - 64]. The HSC megadose was obtained by supplementing BM with G-CSF mobilized PBSCs, both depleted of T lymphocytes. After a highly immunosuppressive and myeloablative regimen which included total body irradiation (TBI) in a single fraction at fast dose-rate, Cyclophosphamide (CY), anti-thymocyte globulins (ATG) and thiotepa, 80% of 36 adults with advanced stage acute leukemia engrafted and although no post-transplant immunosuppressive therapy was administered, the incidence of GvHD was under 10% [65, 66]. Over the years modifications to this approach have led to remarkable progress. Since January 1999, CD34+ cells have been selected in a one-step automated procedure using the Clinimacs device, which yields high numbers of CD34+ cells in a simple and relatively fast procedure [67, 68]. In haploidentical transplants, donor T-cells are extensively depleted, so they cannot exert the GvL-effect, which is then mediated by NK cells which are one of the earliest lymphocyte sub-populations to recover after HSCT. NK cells play a crucial role in the normal innate immune system and are implicated in tumor surveillance since, through recognition of HLA class I antigens, they preferentially kill malignant target cells while sparing normal cells [4, 46].

NK cell alloreactivity exerts not only a GvL effect which prevents leukemia relapse, but also seems to prevent GvHD (by killing host dendritic cells) and reduce infectious mortality [69 - 71]. The ability to access to different types of donors and SC sources, allows physicians to customize the transplant procedure according to age, underlying disease and patient's clinical conditions and makes it a curative option for a wide range of hematological malignancies, non-malignant diseases and solid tumors, as described below. Moreover, advances in genetic and biological analyses, and new targeted-therapies for cancer, have greatly improved patient selection and restricted indications to HSCT so as to obtain the best benefit.

## Stem Cell Transplantation in Hematological Malignancies

### *Acute Myelogenous Leukemia (AML)*

AML occurs in children and adults, but is usually found in middle-aged to older people. Rare before the age of 45, the average age of a patient with AML is 67 years. The American Cancer Society estimates there will be about 20,830 new cases of AML in the United States in 2015, mostly in adults, with about 10,460 deaths (NIH, National Cancer Institute Surveillance and Epidemiology). Sustained leukemia-free survival is achieved in under a third of patients who receive conventional chemotherapy, even though most achieve complete remission. Results are dismal in elderly patients, with cure rates under 5% [20]. Good-risk AML is associated with cytogenetic abnormalities like inv16 or t(8;21), and patients respond well to high-dose cytarabine-based protocols [72, 73]. Patients with acute promyelocytic leukemia (APL), bearing t(15;17), achieve excellent outcomes with regimens containing all-transretinoic acid (ATRA) and arsenic in addition to chemotherapy [74]. In these populations with favorable prognosis, first chemotherapy-induced remission is usually long-term, so allogeneic transplantation does not improve outcome and is generally indicated only if relapse occurs. On the other hand, patients with high-risk AML have poor outcomes with standard induction chemotherapy alone. Sustained survival rates are consistently below 15% [75 - 78]. Poor prognostic factors include unfavorable cytogenetic abnormalities, a high white blood cell count at diagnosis, need for more than one induction cycle to achieve remission, and AML secondary to previous chemotherapy or radiation [79 - 82]. Systematic assessment of co-morbidities is important in determining appropriate treatment, and should be performed routinely in patients with AML because co-morbidities substantially increase the mortality risk with allogeneic transplantation. Although the hematopoietic cell transplantation-specific co-morbidity index was developed as a prognostic tool to estimate risk with transplantation, it has predictive capability for patients undergoing chemotherapy alone for AML [83 - 85]. Allogeneic BMT has been extensively studied in patients with AML as post-remission therapy. Initially administered to patients with refractory disease, 10 to 20% of them enjoyed long-term disease-free survival (DFS). Most studies reported that survival rates after allogeneic transplantation are more than twice as good as with chemotherapy alone [86, 87]. A large prospective study that identified patients as 'high-risk' on the basis of unfavorable cytogenetics or persistent blasts on day 15 of chemotherapy reported survival at 4 years in the majority of patients undergoing allogeneic transplantation from sibling or unrelated donors [88]. Approximately 40% of patients who do not have cytogenetic abnormalities are classified as intermediate risk and choice of post-remission therapy varies widely among clinicians and institutions. In these patients, molecular biology led to the discovery of genetic abnormalities that could not be detected by standard cytogenetic analysis [89]. So far, these include nucleophosmin gene (NPM1) mutations, the FMS-related TK-3 gene due to internal tandem duplications (FLT3-ITD) and the CCAAT/enhancer binding protein α gene (CEPBA). A large retrospective analysis by the German-Austrian AML study group found that allogeneic transplantation provided significant survival benefits only for patients bearing the FLT3-ITD mutation or wild-type NPM1 and CEBPA without FLT3-ITD [89]. Thus, genetic abnormalities that only molecular biology can detect are used to assess prognosis and select treatment.

Since prospective randomized studies demonstrated that allogeneic transplantation was not consistently better than chemotherapy alone for patients with AML in first remission, many patients with AML were referred for transplant beyond first remission or never referred. Only the recent HOVON-SAKK trial achieved high compliance rates (82%) for allogeneic transplantation candidates and demonstrated significantly better survival in transplant recipients. A meta-analysis of 4,000 AML patients reported a significant survival benefit at 4 years post-transplant for patients with poor- or intermediate-risk cytogenetics who had an HLA- identical donor [90]. In patients with advanced disease, allogeneic transplantation provides a lower chance of cure, but is often the only treatment available. Advances in HLA typing and supportive care have improved outcomes, opening up new perspectives for patients with advanced stage AML.

## Myelodisplastic Syndromes (MDS)

Found prevalently in adults, the risk of developing MDS increases with age. MDS is sometimes secondary to chemotherapy and has a higher probability of occurrence after treatment for pediatric acute lymphocytic leukemia (ALL), Hodgkin's disease (HD) or non-Hodgkin's lymphoma (NHL). MDS may involve one or more kinds of blood cell. The number of "blasts" in the bone marrow and blood is a prognostic factor, with more indicating worse prognosis. Prognosis is also affected by abnormal BM cytogenetics, with lack of part of chromosome 5 (5q-syndrome) being the most common. MDS includes an extremely heterogeneous group of myeloid malignancies with different natural courses, risk factors and prognosis. Innovative drugs like hypomethylating agents (HMA, Azacytidine for example) provide valid alternatives to HSCT as therapy [91]. Since most MDS patients are elderly, HSCT can only be offered to a carefully selected subset who need optimal pre- and post- transplant treatment to achieve long-term disease control and at the same time maintain an adequate quality of life. In a decision-analysis study that included patients under 60 years of age, delayed transplantation was associated with maximal life expectancy, with an even more marked survival advantage for patients under 40 [92]. Nevertheless, the introduction of RIC which reduced early transplant related mortality (TRM), significantly increased numbers of MDS patients who undergo HSCT. With better understanding of disease biology and prognosis, different conditioning regimens and diverse graft sources, the transplant strategy should be tailored to the individual candidate to maximize benefits. Together with the FAB and the WHO classifications, prognostic scoring systems in MDS are the most appropriate tools for treatment stratification [93].

The International Prognostic Scoring System (IPSS), takes total scores of three fundamental parameters into consideration: 1) percentage of BM blasts, scored from 0 to 2 (<5%, 5 -10%,11 to 20%, 21- 29%); 2) karyotype, scored from 0 to 1: Good if normal, or bearing the following abnormalities:

-Y, del(5q), del(20q); Poor if complex (3 or more abnormalities) or bearing chromosome 7 anomalies; Intermediate: all others; 3) cytopenia, scored as 0 or 0.5 (defined as Hb <100g/L, ANC < 1,800/uL, and platelets < 100,000/mm3) [94].

The lowest scores have the best outlook. The IPSS categorizes MDS into 4 groups: 1) low risk; 2) intermediate-1 risk (Int-1); 3) intermediate-2 risk (Int-2); and 4) high risk, and only high or Int-2 risk patients are suitable candidates for early allogeneic SCT [93]. On the other hand, the WHO classification-based prognostic scoring (WPSS) incorporated transfusion dependence [94] and identified five risk groups of patients with different survival, allowing for real-time assessment of MDS prognosis and patient selection for HSCT.

Risk category may change over the course of the disease. This can aid in making treatment decisions, particularly in patients with lower-risk MDS, who may have indolent disease (Table **1**). Validated in a study by the Gruppo Italiano Trapianto Midollo Osseo (GITMO) that included 406 patients [95], the WPSS reached prognostic significance on overall survival (OS) and probability of relapse. Its validity was also confirmed in a cohort of 60 Southeast Asian patients [96].

**Table 1. The WHO classification-based Prognostic Scoring System (WPSS) for myelodysplastic syndromes (MDS).**

| Score | | | | |
|---|---|---|---|---|
| **Variable** | **0** | **1** | **2** | **3** |
| WHO category | RA,RARS,del(5q) | RCMD,RCMD-RS | RAEB-1 | RAEB-2 |
| Karyotype | Good | Intermediate | Poor | - |
| Transfusion requirement | No | Regular | - | - |
| WPSS risk group | (0) Very low | | | |
| | (1) Low | | | |
| | (2) Intermediate | | | |
| | (3-4) High | | | |
| | | | | (5-6) Very high |

RA = refractory anemia; RAEB = RA with excess blasts; RARS = RA with ringed sideroblasts; RCMD = refractory cytopenia with multilineage dysplasia; RCMDRS = RCMD with ringed sideroblasts; WHO = World Health Organization.

## Acute Lymphoblastic Leukemia (ALL)

Overall, about 4 in every 10 cases of ALL are in adults while the rest are in children. Children under 5 years of age have the highest risk of developing ALL. The risk then declines slowly until the mid-20s, and begins to rise again slowly after 50 years of age. Although most cases of ALL occur in children, most ALL deaths (about 4/5) occur in adults. Children may do better because childhood and adult ALL are different diseases or because children's bodies can often handle more aggressive treatment than adult's, or some combination of both factors. For 2015 the American Cancer Society estimates about 6,250 new cases of ALL will develop in adults and children (3,100 in males and 3,150 in females), with about 1,450 deaths (800 in males; 650 in females). ALL is a heterogeneous disease, due to expression of different biological and clinical features which determine risk of poor outcome. Currently, the prognostic accuracy of ALL has been improved by minimal residual disease (MDR) monitoring after induction and early consolidation therapy. It should be borne in mind that MRD is also used for deciding on additional treatments to prevent post-transplant relapse. Although allogeneic HSCT is an effective post-remission therapy, one of the main challenges is still patient selection through precise prognostic stratification. Patients with high-risk ALL appear to benefit most from allogeneic HSCT over conventional chemotherapy, though TRM remains significantly higher. Appropriate timing for allogeneic HSCT is crucial, given that the prognosis of relapsed ALL is very poor and the possibility of achieving a second complete remission (CR) is uncertain. Thus, in most patients HLA typing should be performed at diagnosis for considering transplant options early during treatment. The French multi-centre randomized trial LALA-87 found that allogeneic HSCT offered no significant advantage over chemotherapy or autologous SCT (ASCT) to patients with standard-risk ALL [97]. The LALA-94 study by the same group stratified only high-risk patients with donors to allogeneic HSCT and reported the 5-year DFS was 45% in transplanted patients *versus* 23% in patients without donors (p=0.007) [98]. In what was one of the worst forms of leukemia *i.e.* Philadelphia-positive ALL with t(9;22) translocation, administration of TK inhibitors (TKIs) pre- and post-allogeneic HSCT appeared to improve outcomes significantly. In the absence of specifically designed clinical trials, allogeneic HSCT associated with TKIs- therapy remains the most effective post-remission therapy, achieving 88% 3-year event free survival (EFS).

## Chronic Myelogenous Leukemia (CML)

In CML, which accounts for about 10% of all leukemias, a genetic change forming an abnormal gene called BCR-ABL takes place in myeloid cell precursors. Left untreated, CML evolves into a hard to treat, fast-growing acute leukemia. Treatment options for CML, whether in adults or occasionally in children, have changed radically in the past 14 years with the advent of TKIs which have replaced allogeneic HSCT as first-line therapy [99]. Today, predictive models even identify patients who will achieve durable remissions with a second-line TKI after Imatinib failure [100]. The estimated 5-year EFS in these patients is nearly 90%. Long-term follow-ups of TKI-treated patients and availability of new TKIs (dasatinib, nilotinib) with activity against resistant CML, mean allogeneic HSCT can be safely postponed and only offered to patients who are unable to achieve remission or who progress very early to blast crisis.

## Multiple Myeloma (MM)

MM is an adult plasma-cell disease with an estimated 6.3 per 100,000 new cases per year and an almost 50% mortality rate. High-dose chemotherapy followed by ASCT has long been standard frontline consolidative therapy for patients with newly-diagnosed MM. Compared with conventional chemotherapy, CR rates are higher and EFS and OS are better. Response to therapy has long been associated with improved long-term outcomes and, until the advent of new pharmacological agents, only ASCT induced a very good partial response (VGPR) or CR in many patients [101]. A population-based study of 45,595 patients showed significant increases in survival in patients under 80 years old who were treated between 1973 and 2009. Introducing the new anti-angiogenetic drugs (Thalidomide, Lenalidomide) and proteasome inhibitors (Bortezomib) into induction regimens, led to VGPR rates of over 60% even before transplant, with better CR and OS rates, thus challenging the role of ASCT. At present new agent-based therapy combined with tandem ASCT for selected patients seems to provide promising outcomes and is currently under investigation. Allogeneic SCT was tested in patients with MM but, given the high rates of TRM and chronic GvHD, its role remains undefined even though post- transplant consolidation and maintenance strategies have improved progression free survival (PFS) and OS.

# Lymphoma

Lymphomas are divided in Hodgkin's diseases (HD) and non-Hodgkin's lymphomas (NHLs), and treatment options for both include chemo- and radiation therapy. Monoclonal antibody therapy with Rituximab (anti-CD20) in CD20+ NHLs [102], and with Brentuximab (anti-CD30) in HD [103] has drastically changed the natural history of these diseases and led to selected patients who could best benefit from HSCT [102]. ASCT was successful in patients with refractory or relapsed disease.

## Hodgkin's Lymphoma (HL)

HL occurs in all age groups. It is most common in early adulthood (aged 15 to 40), peaking in the 20s, and after 55 years of age. Prognosis is excellent in newly diagnosed patients with advanced stage HL as the vast majority respond well to radio and chemotherapy alone. In contrast, the prognosis of most patients relapsing after first-line therapy is poor. In patients in relapse or in second CR, high-dose chemotherapy and ASCT is treatment of choice. ASCT is also an option for patients with primary refractory disease. Allogeneic HSCT is still considered an experimental procedure for relapsed or refractory HL [104, 105].

## Non-Hodgkin Lymphomas (NHLs)

NHLs are a heterogeneous group of malignancies, accounting for about 4% of all cancers, which may be low, intermediate, or high grade, or even miscellaneous. Like the course of NHLs which varies greatly, from indolent to rapidly fatal, clinical and biological features are very diverse. As new disease markers were discovered over time, classifications changed, prognosis improved and choice of treatment was optimized. Although some types of NHLs are common in children, NHLs can occur at any age, with half the patients being over 66 years of age. Risk increases with age.

## Diffuse Large B-cell Lymphoma (DLBCL)

DLBCL is typical of older individuals, with a median age of approximately 70 years and an annual incidence is about 7-8 cases per 100,000 individuals. DLBCL is a very aggressive tumor, with main risk factor being immune deficiency such as is found in HIV+ patients or in the immediate HSCT post-transplant period. ASCT after second-line chemotherapy is treatment of choice for relapsed or refractory disease [105]. Allogeneic HSCT should only be considered in clinical trials and for highly selected patients.

## Follicular Lymphoma (FL)

FL, a B-cell derived lymphoma, is the most indolent NHL, accounting for approximately 20 to 30% of all NHLs. In its mild form, the "watch and wait" approach is recommended. FL may eventually transform into a more aggressive lymphoma which requires more intensive treatments. The National Comprehensive Cancer Network guideline panel recommended Rituximab-based therapies followed by ASCT for FL in second or third remission, relapse or transformation with good outcomes. Allogeneic HSCT is reserved for patients who have failed, are likely to fail, or are unable to proceed to salvage ASCT, but its late application may not be effective, especially in cases of refractory disease [105].

## Mantle Cell Lymphoma (MCL)

MCL, comprising about 6% of all NHL cases, arises from B-cells within the mantle zone that surrounds normal germinal center follicles. Patients with MCL have a median age of about 60 years and usually present with advanced disease. Allogeneic HSCT has a limited role as consolidation in chemotherapy-sensitive MCL. Given the poor prognosis of relapsing MCL after ASCT and chemotherapy-refractory disease, allogeneic HSCT after a RIC regimen is a reasonable option. Outcomes are, however, rather poor.

## Burkitt's Lymphoma (BL)

BL, a highly aggressive B-cell lymphoma, is endemic in Africa [106], sporadic in other countries, or immunodeficiency associated. It is caused by c-MYC gene translocation and deregulation on chromosome 8, with the most common variant being t(8;14). BL accounts for 2.3% of hematological malignancies and occurs prevalently in children. EBV infection is strongly associated with the disease. Most cases respond well to standard chemotherapy, but recurrent disease can be consolidated or salvaged with ASCT or allogeneic HSCT and outcomes are similar after both approaches. The latest BL protocol by the Associazione Italiana Ematologia Oncologia Pediatrica (AIEOP), in collaboration with the Gustave Roussy Institute of Paris, proposes Rituximab in association with standard chemotherapy in high-risk patients with the aim of reducing the relapse rate and the number of patients needing transplantation as consolidation/salvage treatment.

## Stem Cell Transplantation in Non-neoplastic Hematological Malignancies

### *Severe Aplastic Anemia (SAA)*

Acquired SAA clinical symptoms depend on the degree of cytopenia and on the time required for it to develop. The overall incidence is 2.34 per million people per year and it increases with age [107]. Severe cytopenia may cause fever, bleeding and life-threatening infections. SAA is almost always fatal if untreated with death ensuing from pancytopenia complications, but 85-90% patients survive today for years after diagnosis. The principal therapeutic strategies for SAA are HSCT and immunosuppressive therapy [77, 78]. First-line therapy is a matched sibling un-manipulated BM transplant, preceded by conditioning with CY and ATG (CY 50/mg/kd/die x 4 days with 3 days of ATG). Since SAA affects both HSCs and the BM microenvironment, transplantation outcomes are highly dependent on the number of infused nucleated cells because providing a high number ensures interactions between donor SCs and the recipient BM niche. The recommended BM dose is $4x10^6$/Kg mononuclear cells [108 - 113]. Transpla-ntation is followed by CSA and short-term MTX as prophylaxis for GvHD [114]. As best outcomes are observed in young patients, this is still standard of care when a matched sibling is available. Current survival rates are 91% in children <16 years and 74% in patients over 16 years. Survival rates drop, due to a greater incidence of chronic GvHD, if matched sibling PB is used rather than BM (from 85% to 73% in patients under 20 years; from 64% to 52% in patients over 20 years) [108 - 113]. Age impacts strongly on outcomes: the actuarial 10-year survival after BMT in the last decade is, respectively, 83%, 73%, 68% and 51% for ages 1 to 20, 21 to 30, 31 to 40 and 40 and older. For this reason, the current first-line therapy for patients with SAA and age > 40 years is immunosuppressive therapy alone. A sibling BMT with low-dose CY, fludarabine and ATG may be considered as second-line treatment [108 - 113]. A MUD transplant is still second-line treatment for patients without sibling donors. The Japanese group and the European Bone Marrow Transplantation Group (EBMT) are currently testing a conditioning regimen with Fludarabine-CY-ATG and low-dose TBI (2 Gy) followed by a MUD graft [109]. Because of high rejection rates and low cell numbers, UCBT are not recommended for SAA patients. However, a recent novel approach co-infuses CB with haploidentical CD34+ cells to aid engraftment. A small study presented at the 2012 ASH Meeting, reported that early T-cell engraftment was of CB origin; myeloid engraftment was initially from the haploidentical SCs; dual engraftment CB and haploidentical cells followed and then loss of haploidentical cells with full donor myeloid engraftment from the CB cells [115]. In T-replete haploidentical transplantation for SAA, a novel approach proposes non-myeloablative conditioning with high-dose CY given after transplantation to prevent GVHD [115]. The rationale is that it would deplete proliferating donor alloreactive T cells and spare quiescent non-alloreactive T cells. In SAA post-transplant chimerism monitoring is of crucial importance. Persistence of up to 20% host-chimerism is expected and does not necessarily indicate SAA relapse. If it rises, whether symptoms are present or not, immunosuppressive therapy should be reintroduced if suspended, or the patient should be given donor lymphocyte infusions.

**Inherited Bone Marrow Failure Syndromes (IBMFS)**

Inherited bone marrow failure syndromes (IBMFS), a heterogeneous group of rare hematological disorders, are characterized by poor hematopoiesis and a predisposition to cancer, with clonal evolution towards MDS or AML [116]. The most commonly associated congenital abnormalities are DNA repair failure, telomere deregulation and mutations affecting ribosome assembly or function. HSCT is treatment of choice for IBMFS and in the absence of a suitable donor, the only other option is supportive care with blood products and antibiotics [116]. Fanconi's Anemia (FA), with an estimated incidence of 1: 360,000 births, is characterized by congenital organ and skeletal abnormalities, progressive BM failure and a predisposition to develop cancer because of DNA repair failure [116]. HSCT, the only curative option, increases the risk of secondary cancers in a population that is already at high risk [116, 117]. The optimal timing of HSCT is difficult to establish. Ideally, patients should be transplanted before they receive 20 red cell and/or platelet transfusions, and/or androgen therapy [118] or develop persistent or moderately severe cytopenia or clonal evolution. To limit toxicity and TRM, a RIC regimen should be used with low-dose CY and fludarabine. To prevent further impairment of DNA repair, low-dose TBI should be administered only in UCBT and/or when risks of rejection or relapse are high. Prognosis is generally poor especially for patients who are transplanted with clonal evolution [117].

Clinically, Dyskeratosis Congenita, a very rare IBMF with an estimated prevalence of 1/1,000,000, is characterized by reticulated skin hyper pigmentation, nail dystrophy and oral leukoplakia and clonal evolution. Genetically, telomerase dysfunction and ribosome deficiency are typical and mutations were identified in eight genes (DKC1, TERT, TERC, TINF1, NOP10, NPH2, TCAB1 and RTEL1) involved in the telomerase complex. In 80% of patients BM failure is the main indication to HSCT [116, 119] and non-myeloablative conditioning regimens containing Fludarabine are recommended to prevent toxicity and TRM. Matched sibling donors remain donors of choice, as mismatched related or MUD transplants are associated with poorer outcomes [116, 119].

Severe congenital neutropenia (SCN) is characterized by early diagnosis of profound neutropenia which predisposes the patient to recurrent severe and life-threatening bacterial infections [119]. Estimated incidence is 1 in 200,000 individuals. SCN is most frequently caused by mutations in the ELANE and SBDS genes, but mutations in HAX1, G6PC3, WASP, VPS45, GATA2 and GFI1 genes were also observed. GATA2 mutations were identified as cause of mild, congenital neutropenia associated with a high risk of leukemic transformation which is usually refractory to chemotherapy [119]. Indications to HSCT include recurrent bacterial infections; G-CSF resistance (> 20 µg/kg/day for more than 1 month without normalizing neutrophil counts) even without infection or clonal evolution; long-term therapy with high-dose G-CSF, because of the potential, but not proven risk of clonal evolution. Most patients with SCN received a non-myeloablative conditioning regimen usually with busulfan and CY. Prognosis is often favorable with good outcomes after transplantation.

Another form of congenital neutropenia is the very rare Shwachman-Diamond syndrome. A recessive disorder, it is characterized by exocrine pancreatic insufficiency, skeletal abnormalities, mild intellectual retardation and cytopenia [120]. HSCT is indicated when cytopenia worsens, increasing transfusion dependence and/or clonal evolution. A RIC regimen seems best to prevent toxicity and TRM. Outcomes are uncertain as few cases have been reported. Blackfan-Diamond's anemia, an autosomal dominant IBMFS, affects approximately 5 to 7 per million liveborn infants worldwide. It is characterized by lack of erythroblast precursors, leading to macrocytic anemia and sometimes low white blood cell and platelet counts. Indications to HSCT, the only curative option in transfusion-dependent patients, are non-response to steroids with no change in reticulocyte counts after two cycles of steroid therapy (1mg/kg/day); clonal evolution and aplastic anemia [116, 119]. Only matched sibling donors are recommended. A c-MPL gene mutation leading to a defective thrombopoietin receptor, underlies congenital amegakaryocytic thrombocytopenia *i.e.* lack of megakaryocytes in BM. The exact prevalence in unknown and less than 100 cases have been reported in literature. As in other IBMFS, HSCT is the only option and indications are severe isolated thrombocytopenia or pancytopenia and clonal evolution. The standard myeloablative approach based on fludarabine and either busulfan or treosulfan is recommended for clonal evolution [116, 119].

A reduced-intensity approach may be considered for patients with pancytopenia. Prognosis is good, particularly if patients are transplanted before clonal evolution since otherwise risk of relapse is high.

Paroxysmal Nocturnal Hemoglobinuria (PNH), a rare acquired hematopoietic stem cell disorder with a worldwide estimated prevalence of 1-5 cases per million, is characterized by complement system-mediated hemolytic anemia, marrow failure or venous thrombotic events. In PNH a somatic mutation in the X-linked, phosphatidylinositol glycan class A (PIG-A) gene leads to deficient expression of PIG-A-anchored proteins on red blood cells [121]. Today, treatment of choice is Eculizumab, a targeted monoclonal antibody against the C5 complement fraction which interferes with complement-mediated cell lysis. HSCT is now reserved only for non-responding patients and to reduce risk of chronic GvHD, BM is the best source of SCs. There are no clear indications about the conditioning regimen as no difference emerged in survival rates after a myeloablative or a reduced intensity regimen [122].

The congenital dyserythropoietic anemias (CDAs) are a hetereogeneous group of clinically challenging but biologically fascinating rare inborn disorders, with an unknown prevalence and less than 100 cases reported in literature. Unlike other IBMFS they are characterized by morphological erythroblast abnormalities that lead to ineffective erythropoiesis. Other hematopoietic lineages seem to be unaffected. Indications to HSCT are splenectomy failure and transfusion-related iron overload. Outcomes are best with HLA identical donors [123].

**Hemoglobin Diseases**

Due to genetic defects in hemoglobin protein-chains, Thalassemia major, originating in Mediterranean, Middle Eastern and Asian countries, and sickle cell anemia (SCA), originating in Africa, are the two most widespread hereditary hemoglobin pathologies in the world. Both spread globally through migration. For Thalassemia major, the annual incidence of symptomatic cases is estimated at 1/100,000 worldwide, and 1/10,000 in Europe. SCA is much more common in certain ethnic groups, as it is caused by a genetically evolved hemoglobin to protect individuals from malaria. For this reason, SCA is more frequent in those geographic areas where malaria is endemic. Among African Americans, approximately one person out of every 500 is affected by SCA and about 8% has sickle cell trait. Standard treatment for both these hemoglobin diseases is blood transfusions and iron chelating. Modern drugs, *e.g.* deferasirox, are very efficacious and have reduced the number of patients requiring HSCT. However, allogeneic HSCT from an HLA identical sibling, with MUD HSCT as a viable alternative, offers the only cure as SCs correct the basic genetic defect by replanting genes that are essential for normal hematopoiesis. If a donor is available, HSCT should be performed as soon as possible as good outcomes range from 80-88%, but worsen with more advanced disease and organ complications that are mainly due to prolonged exposure to iron overload [124]. For thalassemia, the standard conditioning regimen contains CY and busulfan. Hepatomegaly of more 2 cm, portal fibrosis and irregular chelation history were associated with poor outcome. The Lucarelli group recently reported the outcomes of 31 children with thalassemia who received transplants from haploidentical donors (27 mothers, 2 brothers and 2 fathers). All patients received CSA for GVHD prophylaxis for the first 2 months post-transplant. Although all patients showed full donor chimerism by day 14, seven rejected their grafts and survived with thalassemia. Two patients died of TRM. Transplantation was successful with complete allogeneic reconstitution in 19 cases and with mixed chimerism in 3. All 22 cured children are no longer transfusion- dependent, with hemoglobin levels ranging from 10.3 g/dL to 13.8 g/dL [125].

In children with SCD outcomes of HSCT from matched sibling donors are excellent. Transplantation should be performed before the development of irreversible sickle vasculopathy-related complications and has been offered primarily to younger patients with overt symptoms. Timing and indications to transplantation also depend on the effects of treatments such as hydroxyurea and blood transfusions and on the impact of transplantation on organ damage. Although the results of HSCT after myeloablative conditioning are encouraging, with DFS of 85% and 10% TRM, a RIC regimen is recommended to reduce TRM. Most patients receive 14–16 mg/kg busulfan and 200 mg/kg CY. Additional immunosuppressive agents sometimes include equine or rabbit ATG or total lymphoid irradiation. CSA, alone or in combination with methylprednisolone or MTX is usually administered post-transplant as GVHD prophylaxis. Use of UCBT is limited in patients with SCD, and appears to be associated with a high risk of graft rejection and GvHD [126].

**Stem Cell Transplantation in Solid Tumors**

Prior to the introduction of high-dose chemotherapy (HDC) with autologous SC rescue, marrow tolerance was the limiting factor in chemotherapy escalation for the treatment of malignancies. More intense treatment of certain malignancies, in particular in pediatric and adolescent populations, became feasible with the ability to safely harvest, store and re-infuse a patient's own HSCs. Doses of cytotoxic therapies for cancer could safely proceed beyond marrow tolerance, as transplanted autologous cells would soon re-build the hematopoietic system. With myeloablative HDC, no hematopoietic recovery can occur without the stored autologous HSCs; while with sub-myeloablative HDC regimens, SC rescue is used to speed recovery, decrease toxicity and shorten the interval between courses of chemotherapy [127 - 129]. Indications for use of HDC with SC rescue include: a) a tumor with good response to induction chemotherapy, but poor 3 or 5-year EFS; and (b) a HDC regimen that can utilize multiple active agents against the disease, particularly if they were not used during induction therapy [129]. The only diseases that satisfy these criteria with clinical trials indicating better outcomes are, at present, germ cell tumors, Ewing or Ewing-like sarcomas, and high-risk neuroblastoma. Attempts to treat osteosarcoma and rhabdomiosarcoma with autologous and allogeneic transplantation have provided encouraging, but not brilliant results in the long-term [130 - 134]. The advent of innovative targeted therapies like monoclonal antibodies and TKIs, has made HSCT useless and too risky in recurrent and/or metastatic malignancies such as breast and renal cell cancer [135, 136].

## Germ Cell Tumors

Germ cell tumors are a heterogeneous group of diseases for age of onset and clinical features. They are separated into two groups: gonadal tumors of the testis and ovary, or extra-gonadic with the most typical localizations being sacrococcygeal, retroperitoneal, and mediastinal. They account for 2-3% of malignancies in children and prognosis is often favorable.

Metastatic germ cell tumors in males and females are potentially curable with cisplatin-based combination chemotherapy with or without associated surgery [137 - 139]. For patients who relapse or whose first- line therapy fails, second-line conventional chemotherapy (cisplatin and ifosfamide combinations) is associated with only 15-25% successful salvage [138 - 142]. One study retrospectively reviewed the outcome of 21 patients with relapsed germ cell neoplasm (testis, ovary, or extra-gonad) over a period of 11 years, who received HDC (carboplatin and etoposide) followed by ASCT. HDC/ASCT provided significant long-term EFS as salvage therapy for both male and female patients. The main risk factor was cisplatin-refractoriness or resistance during first-line treatment. Patients in stable or progressive disease after first-line therapy are defined platinum-refractory; patients with partial response (PR) or residual tumor or in CR who relapse < 6 months after induction treatment are defined as platinum-resistant;

patients in CR who relapse after > 6 months are defined as platinum-sensitive. EFS at 5 years was better in patients who are not refractory to cisplatin (77%), while those who have cisplatin-refractory disease can expect only a 10-20% chance of durable remission [138, 141, 142].

## Ewing and Ewing-like Sarcomas

Ewing sarcoma, the second most common bone tumor in children after osteosarcoma, has an incidence of 3 per million in patients under 20 years of age. In 90% of cases, patients are between 5 and 25 years of age with 65% of cases being diagnosed between 10 and 20 years of age, 25% before 10 years of age and about 10% over 20 years of age. Therapy includes surgical resection, anthracycline and alkylator chemotherapy (typically doxorubicin and ifosfamide) and in some cases irradiation. Outcomes are much worse in patients with metastases (4-year OS 39%) and relapse (10-year OS 10%) [140, 143, 145]. In patients with metastases, escalating core treatment agents (ifosfamide, doxorubicin and CY) or adding ifosfamide/etoposide to a standard regimen only seemed to increase short term toxicity and secondary MDS [146]. In 36 patients with large, unresectable tumors and stage 4 patients with metastases to the BM, a study from Meyers *et al.* investigated the efficacy of melphalan, etoposide and TBI followed by ASCT. It led to no improvement in 2-year survival (20%) over conventional therapy [147]. Using allogeneic and autologous SC sources, three other studies replicated these disappointing results and high TRM rates, raising further concerns [148, 149]. The Italian and Scandinavian Sarcoma Groups designed a joint study (ISG/SSG IV) to improve the prognosis for patients affected by Ewing's family tumors and synchronous metastatic disease limited to the lungs, the pleura, or a single bone [150]. The protocol consisted of intensive five-drug combination chemotherapy, surgery and/or irradiation as local therapy and high-dose busulphan/melphalan plus autologous SC rescue and total lung irradiation as consolidation treatment [150]. This intensive approach was feasible and long-term survival was achieved in nearly 50% of patients, with 40% EFS at 120 months. In two single center studies, a total of 32 relapsed Ewing sarcoma patients underwent HDC/HSCT (26 autologous; 6 allogeneic) with only 15 being reported as long-term survivors, thus constituting a rare subgroup that achieved a second remission [147, 150, 152]. It is unclear whether HSCT provided an advantage over intensified standard chemotherapy. In the presence of gross residual disease outcomes after HDC/ASCT are exceptionally poor (5-year OS 19%) [153]. AIEOP is promoting a protocol for patients with very poor prognosis (refractory, relapsed or stage IV disease with BM involvement), who received up-front chemotherapy with Cisplatin 120 mg/sm for two cycles, followed by 8 cycles according to the ISG/SSG IV protocol, and then allogeneic sibling-matched HSCT to exploit the GvT effect of the donor's immune system. No results are as yet available as this study is still on-going. For non-metastatic Ewing's family tumors, response to chemotherapy is the main factor determining outcome. The ISG/SSG III study identified "good" or "poor" responder patients (GR and PR, respectively) according to necrosis grade post-chemotherapy. After three cycles of up-front therapy, patients were locally treated with surgery and/or radiotherapy and, if poor responders, received HD busulphan/melphalan followed by ASCT. This approach identified patients who would benefit from salvage therapy. At a median follow-up of 64 months (21-116 months), 5-year OS and EFS were 75% and 69%, respectively. Five-year EFS was 75% for GR, 72% for PR treated with HDT and 33% for PR who did not receive HDC/ASCT [154].

**Neuroblastoma**

High-risk neuroblastoma (NB) remains among the most challenging of pediatric malignancies, with limited progress in clinical trials over the past three decades [155, 156]. NB, the third most common tumor in childhood, is the most frequent malignancy in children under 5 years of age. It accounts for 7-10% of all tumors in patients < 15 years of age, with an incidence of 8 per million in people < 16 years of age. Median age at diagnosis is 2 years and 90% of cases are diagnosed under the age of 6. High-risk children over the age of one have either disseminated disease (INSS stage 4: about 40 - 50% of all NBs) or INSS stages 2 and 3 with MycN proto-oncogene amplification (about 3% of all NBs) which is found in 10% -20% of children with stage 3 and occasionally in patients with stage 2. This definition also includes infants (< one year of age) with MycN amplified tumors. See Table **2** for NB staging. Since MycN proto-oncogene amplification was clearly associated with a greater risk of relapse and death from disease progression, patients bearing it may benefit from very aggressive treatment. Pending development of more targeted approaches to improve OS, patients at present receive intense cytotoxic therapies and consolidation with myeloablative regimens followed by the infusion of ASCT. In a pilot study, the Children's Oncology Group (COG) assessed the feasibility and toxicity of a tandem myeloablative regimen: first thiotepa and CY followed by autologous infusion of CD34+ selected peripheral blood stem cells and second carboplatin/etoposide/melphalan (CEM), without TBI, which was again followed by ASCT. This approach was feasible and improved outcomes [156]. Despite the proven efficacy of immunotherapy [157] and the promise of exciting new targeted agents (TKIs, for example), cytotoxic therapeutic intensity and myeloablation followed by ASCT rescue remain fundamental in the treatment of high-risk NB. The ongoing SIOPEN/AIEOP study HR-NBL-1 uses ASCT as consolidation treatment after induction chemotherapy. Patients were randomized to receive Bu/Mel (standard arm) or CEM (actually closed) as conditioning. The second randomization point of this protocol was immunotherapy (anti-GD2 monoclonal antibody plus 13-cis retinoic acid) with or without IL-2. The second randomization is still ongoing. Despite all these approaches and intensive treatments, long-term survival rates for high risk patients with NB is under 60%. While ASCT is well established as consolidation therapy for NB, the role of allogeneic SCT is still controversial.

The Center for International Blood and Marrow Transplant Research (CIBMTR) conducted retrospective analyses of 143 patients with NB who received allogeneic SCT from 1990 to 2007. Patients were divided into two groups: those who had not (Group 1) and those who had (Group 2) undergone ASCT (n=46 and 97, respectively). One-year and 5 -year OS rates were 59% and 29% for Group 1, and 50% and 7% for Group 2. One- and 3- year (but not 5 year-) DFS and OS were significantly lower after transplantation from unrelated donors because TRM was higher in these patients. Patients who were in CR or VGPR at transplant had lower relapse rates and better EFS and OS than those who were not in remission. These results indicated that allogeneic SCT can cure some NB patients. It lowered relapse rates and improved survival in patients without previous ASCT. Therefore, allogeneic SCT after a prior ASCT appears to offer minimal benefit and relapse remains the most common cause of treatment failure [158].

**Table 2. The International Neuroblastoma Staging System (INSS) [151, 160].**

| Stage 1 | localized tumor removed radically, with or without microscopic residual; "representative" ipsilateral lymph nodes, histologically negative (lymph nodes adherent to the tumor removed along with the tumor can be histologically positive, but result does not change the stage). |
|---|---|
| Stage 2 A | localized tumor removed incompletely; "representative" ipsilateral lymph nodes not adherent to the tumor histologically negative for tumor infiltration. |
| Stage 2 B | localized tumor removed completely or incompletely, with ipsilateral lymph nodes not belonging to the tumor positive for tumor infiltration. Any contralateral adenopathy histologically negative for infiltration. |
| Stage 3 | unresectable unilateral tumor infiltrating the midline with or without regional lymph nodes; or localized unilateral tumor with contralateral regional lymph node involvement; or midline tumor with bilateral extension by infiltration or lymph node involvement [159]. |
| Stage 4 | any primary tumor with dissemination and distant lymph nodes, bones, bone marrow, liver, skin and/or other organs (except as described in stage 4S) [159]. |
| Stage 4 S | primary tumor localized (as for stages 1 and 2A or 2B) with dissemination limited to skin, liver and / or bone marrow [159]. |

## HSCT in Immunedeficiencies, Inherited Metabolic Disorders and Autoimmune Diseases

### Transplantation in Immunedeficiencies

Immunedeficiencies (ID) are inherited disorders that lead to severe impairment of innate and adaptive immune systems and early death from infectious complications. Table **3** summarizes the most frequent ID with their associated genetic abnormalities, indicating which can be cured with HSCT. OS rates are over 90% [160 - 162] because in ID donor HSC regenerate healthy immune system cell populations, as shown in Fig. (**4**).

**Table 3. ID, associated genetic abnormalities, immunity defects and HSCT indications.**

| SCID | Genetic Abnormality/Defect | HSCT Indication |
|---|---|---|
| X-linked SCID | γ chain IL-2 receptor | YES |
| Omenn Syndrome | Multiple mutations; unknown mutations in the severe form | YES |
| Di George Syndrome | T-cell immune deficiency | YES |
| Wiskott-Aldrich Syndrome | X-linked | YES |
| Ig deficiency | X-linked, VCI | NO |

*(Table 3) contd.....*

| SCID | Genetic Abnormality/Defect | HSCT Indication |
|------|----------------------------|-----------------|
| **Complement deficiency** | X-linked, lectin PW def | NO |
| **Neutrophil deficiency** | CGD, iper-IgE syndrome | YES in SCN |

VCI=variable common immune deficiency; PW def=pathway deficiency; CGD=chronic granulomatosis disease; SCN=severe congenital neutropenia.

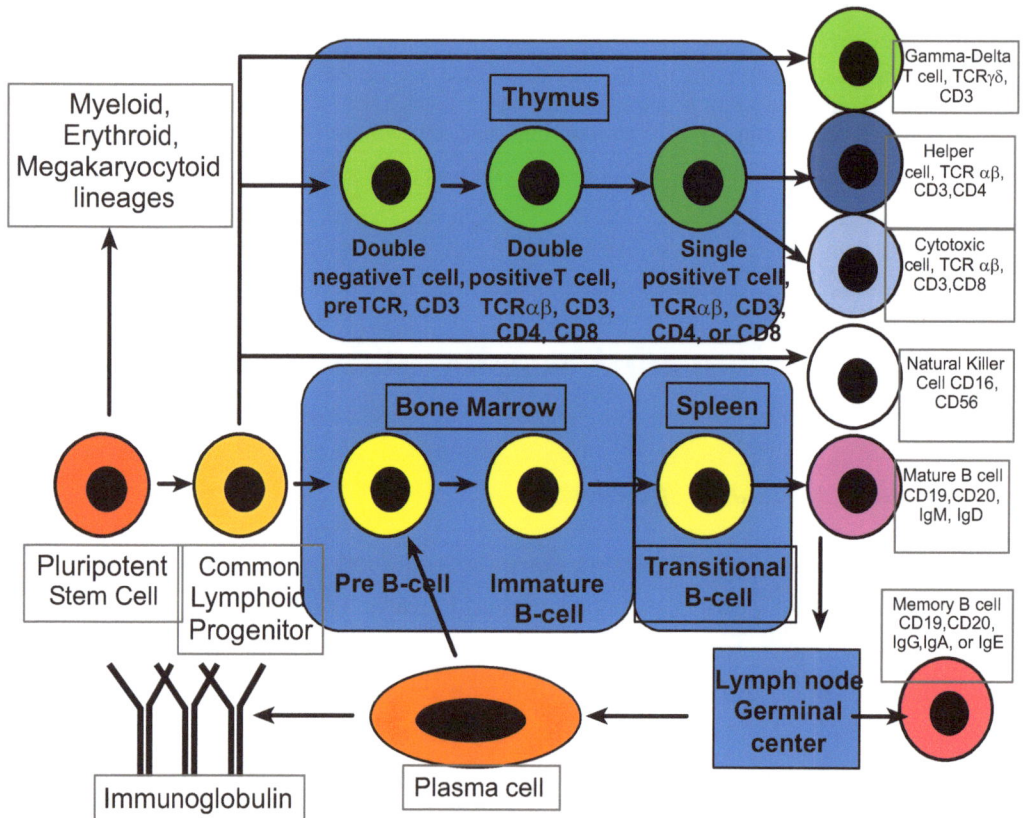

**Fig. (4).** Schema of lymphoid stem cell progenitor regenerates the immune system and differentiates into committed cell lineages in the thymus and peripheral lymphatic tissue.

Severe combined immune deficiency (SCID), the most common ID with an incidence of 1:50-100,000 newborns, is characterized by lack of antigen-specific T-cell and B-cell immune responses combined with severe T-cell lymphocytopenia. HSCT was first performed in SCID patients and all types of allogeneic HSCT fully correct T-cell and, less consistently, B-cell deficiencies. SCID are also one of the best disease for gene therapy because many underlying gene abnormalities have been identified, *e.g.* IL-2 gamma-chain receptor, or defective adenosine- deaminase (ADA). The advantages of gene therapy are B-cell compartment reconstitution and immunoglobulin (Ig) infusion independence, and good functional T- and NK cell repopulation. In 2002, 4/5 patients with X-linked SCID who were safely treated with gene therapy achieved long-term immune system reconstitution [163]. Unfortunately, 2/4 patients developed acute leukemia which was caused by the LMO2 oncogene activation. Interestingly, the LMO2 gene is located beside the retrovirus insertion point and is involved in the onset of T-ALL [164]. Even though on-going gene therapy attempts are using other vectors which are hoped to be safer, HSCT has become once again the only curative option for SCID. HSCT outcomes are influenced by patient age and clinical conditions, B-cell reconstitution, donor availability, and time-lapse to transplantation. Survival is best if the donor is a matched sibling, because of the need for less immune suppression and faster immune reconstitution [165, 166]. Since few SCID patients have a matched sibling, alternative HSC sources are haploidentical donors or UCB. 5-year OS was the same, but B-cell reconstitution, and consequently Ig infusion discontinuation, was better after UCBT [167]. These data were confirmed by a retrospective analysis of 240 transplanted infants by the Primary Immune Deficiency Treatment Consortium, showing survival at 5 years, freedom from immunoglobulin substitution, and CD3+ T-cell and IgA recovery were more better after matched sibling donors than alternative donors. Survival was best, after haploidentical T-cell-depleted transplants with no conditioning regimen in actively infected infants without a matched sibling donor. On the other hand, although RIC or myeloablative conditioning regimens were associated with a CD3+ T-cell count of >1,000/cmm, freedom from immunoglobulin substitution, and IgA recovery, they did not significantly affect CD4+ T-cell recovery or phytohemagglutinin (PHA)-induced T-cell proliferation. Recognition of the specific molecular defect may determine use or not of a conditioning regimen before HSCT for SCID patients [168]. Patients with ADA deficient SCID will develop adequate immunological reconstitution after a matched sibling HSCT, with no need for a conditioning regimen. Those with recombinant activating gene (RAG) deficiency will require chemotherapy conditioning to achieve engraftment and immunological reconstitution.

## Transplantation in Inborn Metabolic Errors

Inborn metabolic errors (IMEs) arise from genetic defects, leading to lysosomal enzyme deficiencies and peroxisomal abnormalities which lead to fat accumulation in vital organs, since peroxisomes are primarily involved in lipid metabolism. Each IME is very rare, but all together they constitute quite a large disease entity. IMEs affect bones, growth, cardiopulmonary status, airways, hearing and vision, neurological and cognitive functions. Onset is in infancy or early childhood with rapid deterioration and early death [170]. IMEs are usually caused by defects in specific hydrolytic lysosomal enzymes that catalyse the degradation of specific substrates [169].

Defective lysosomal enzyme production leads to lysosomal storage diseases (LSDs) which include Mucopolysaccharidosis type-1, Hurler phenotype (MPS-1H) and the inherited lysosomal leucodystrophies, in particular metachromatic leucodystrophy (MLD), X-linked adrenoleucodystrophy (X-ALD) and globoid cell leucodystrophy (GLD) or Krabbe disease [169]. Pharmacological intravenous enzyme replacement therapy is today available for MPS-1, -II, -VI, Gaucher and Fabry diseases, and substrate deprivation therapy, in which small molecules inhibit storage material production is under development. Gene therapy trials are accruing patients with X-ALD and MLD [169]. Allogeneic HSCT is the standard of care in a selected group of disorders like MPS-1 and early X-ALD and, to date, over 2,000 patients with an IEM have successfully been transplanted with survival rates as high as 90%. Despite these successes, residual disease burden remains a major limitation [169].

### Autoimmune Diseases

Autoimmune diseases affect about 5% of the population. Current systemic therapies do not offer long-term efficacy, but targeted therapies, often based on monoclonal antibodies, like Rituximab, are successful and have limited applications of HSCT. Indications to ASCT are available in selected cases of advanced stage multiple sclerosis (MS) or systemic sclerosis (SSc), systemic lupus, Crohn's disease, type I diabetes (T1D), and juvenile idiopathic arthritis [165]. PFS after ASCT is in the range of approximately 50%. In 65 individuals with new-onset T1D, ASCT after a non-myeloablative (ATG +CY) conditioning regimen was successful in 59% who achieved insulin independence within the first 6 months and 32% remained insulin independent after 4 years [170, 171]. A "resetting" of the immune system is hypothesized to occur after ASCT. Since 1996, more than 1,300 patients with autoimmune diseases have been registered by the EBMT and almost 500 patients by the CIBMTR. According to last EMBT guidelines, ASCT or, in patients who have SC donors, allogeneic HSCT should be considered. Patients who relapse after transplantation procedure, can still be treated with standard or biological drugs to which they used to be resistant earlier [45].

## CONCLUSION AND FUTURE PERSPECTIVES

This chapter reviewed in depth the therapeutic use of HSCT from various donor sources in a wide range of malignant and non-malignant diseases. Although transplanted HSC give rise to new haemopoietic and immune systems, their use is by no means limited to hematological diseases. Today, even though main causes of HSCT failure are relapse, GvHD and infectious mortality, HSCT is safe, available in many centers worldwide and a feasible option for patients of most age groups – from the very young to the middle-aged and elderly. The modalities of autologous, matched sibling, MUD and UCB protocols are well established and after matched HSCT, prolonged immune suppression is needed to prevent or control GvHD which is caused by the high number of alloreactive T cells in the graft [43]. The corollary of immune suppression is risk of relapse and life-threatening viral and fungal infections in the first 3-6 months post-HSCT [43]. Most changes and advances have occurred in the haploidentical setting, mainly because the haploidentical family member is usually the most suitable donor to be found and SC selection techniques have become more and more automated. With the T-cell-depleted haploidentical transplant the benefits of NK cell alloreactivity were observed [69 - 71] and antigen-specific adoptive immune therapies were successfully attempted in the absence of pharmacological prophylaxis for GvHD [172]. Since however, immune recovery was very slow, attention focused on negative SC selection with infusion of gamma-delta T-cells [173, 174], or co-infusion of regulatory and conventional T-cells [175, 176]. Given their immune modulatory properties, these T-cell subpopulations facilitated engraftment, controlled leukemia relapse and infectious mortality, without causing GvHD (Fig. **5**).

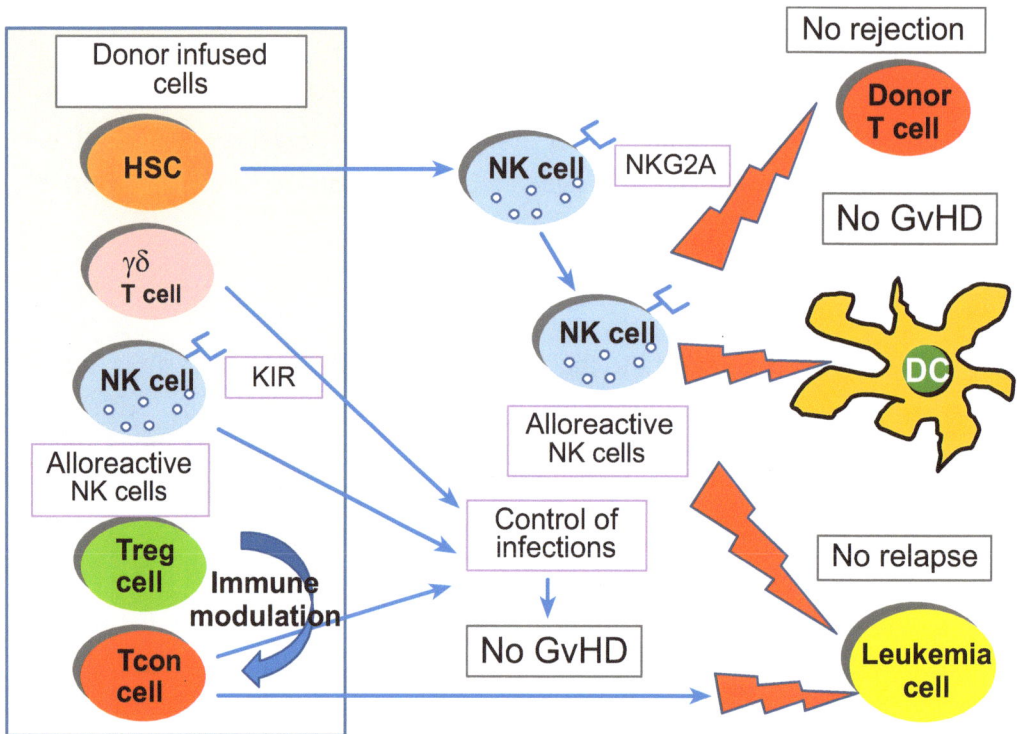

**Fig. (5).** Haploidentical HSCT for acute leukemia. Innovative adoptive cell strategies to improve immune reconstitution and consequently to reduce infection-related TRM and leukemia relapse. DC=dendritic cell.

Interest in T cell-replete full haplotype mismatched HSCT was reawakened by new transplant strategies for GvHD prophylaxis, such as G-CSF–primed grafts [178], post-transplantation rapamycin [178], or high-dose CY in combination with other immunosuppressive agents [179, 180]. T cell-replete HSCT presents two major challenges. Although the high T-cell content of the graft potentially enhances the GVL effect, the T cells induce significant GvHD related morbidity and mortality. However, when using other strategies to prevent GvHD in T cell–replete HSCT, such as administration of high- dose CY after transplantation, a high incidence of leukemia relapse emerged as a major problem. Therefore, efforts are still being made to reduce the incidence and severity of GvHD and to prevent leukemia relapse. One final issue that still remains open is whether targeted or gene therapies will replace HSCT in the future. They are already being used to cure several hematological malignancies and solid tumors and if research in this field continues in its successful way, the HSCT may be reserved only for very highly selected patients who are refractory to all other approaches.

## CONFLICT OF INTEREST

The author confirms that the author has no conflict of interest to declare for this publication.

## ACKNOWLEDGEMENTS

We acknowledge Dr. Seraldine Annu Boyd for editorial assistance.

## REFERENCES

[1]    Lazarus HM, Laughlin MJ. Allogeneic Stem Cell Transplantation. New York: Springer 2003.

[2]    Peccatori J, Ciceri F. Allogeneic stem cell transplantation for acute myeloid leukemia. Haematologica 2010; 95(6): 857-9.
[http://dx.doi.org/10.3324/haematol.2010.023184] [PMID: 20513804]

[3]    Aversa F, Velardi A, Tabilio A, Reisner Y, Martelli MF. Haploidentical stem cell transplantation in leukemia. Blood Rev 2001; 15(3): 111-9.
[http://dx.doi.org/10.1054/blre.2001.0157] [PMID: 11735159]

[4]    Ruggeri L, Capanni M, Mancusi A, Aversa F, Martelli MF, Velardi A. Natural killer cells as a therapeutic tool in mismatched transplantation. Best Pract Res Clin Haematol 2004; 17(3): 427-38.
[http://dx.doi.org/10.1016/j.beha.2004.05.010] [PMID: 15498714]

[5]    Yi BA, Mummery CL, Chien KR. Direct cardiomyocyte reprogramming: a new direction for cardiovascular regenerative medicine. Cold Spring Harb Perspect Med 2013; 3(9): a014050.
[http://dx.doi.org/10.1101/cshperspect.a014050] [PMID: 24003244]

[6]    Witkowska M, Smolewska E, Smolewski P. Hematopoietic stem cell transplantation in children with autoimmune connective tissue diseases. Arch Immunol Ther Exp (Warsz) 2014; 62(4): 319-27.
[http://dx.doi.org/10.1007/s00005-014-0279-9] [PMID: 24604327]

[7]    Heissig B, Dhahri D, Eiamboonsert S, *et al.* Role of mesenchymal stem cell-derived fibrinolytic factor in tissue regeneration and cancer progression. Cell Mol Life Sci 2015; 72(24): 4759-70.
[http://dx.doi.org/10.1007/s00018-015-2035-7] [PMID: 26350342]

[8]    Sureda A, Bader P, Cesaro S, *et al.* Indications for allo- and auto-SCT for haematological diseases, solid tumours and immune disorders: current practice in Europe, 2015. Bone Marrow Transplant 2015; 50(8): 1037-56.
[http://dx.doi.org/10.1038/bmt.2015.6] [PMID: 25798672]

[9]    Passweg JR, Baldomero H, Bader P, *et al.* Hematopoietic SCT in Europe 2013: recent trends in the use

of alternative donors showing more haploidentical donors but fewer cord blood transplants. Bone Marrow Transplant 2015; 50(4): 476-82.
[http://dx.doi.org/10.1038/bmt.2014.312] [PMID: 25642761]

[10]    Weissman IL. Stem cells: units of development, units of regeneration, and units in evolution. Cell 2000; 100(1): 157-68.
[http://dx.doi.org/10.1016/S0092-8674(00)81692-X] [PMID: 10647940]

[11]    Becker AJ, McCULLOCH EA, Till JE. Cytological demonstration of the clonal nature of spleen colonies derived from transplanted mouse marrow cells. Nature 1963; 197(4866): 452-4.
[http://dx.doi.org/10.1038/197452a0] [PMID: 13970094]

[12]    Krause DS, Fackler MJ, Civin CI, May WS. CD34: structure, biology, and clinical utility. Blood 1996; 87(1): 1-13.
[PMID: 8547630]

[13]    Peled A, Kollet O, Ponomaryov T, *et al.* The chemokine SDF-1 activates the integrins LFA-1, VLA-4, and VLA-5 on immature human CD34(+) cells: role in transendothelial/stromal migration and engraftment of NOD/SCID mice. Blood 2000; 95(11): 3289-96.
[PMID: 10828007]

[14]    Stem cell mobilization: methods and protocols, e-Book edited by MG Kolonin and PJ Simmons, Humana Press/Springer, c2012.

[15]    Lapidot T, Petit I. Current understanding of stem cell mobilization: the roles of chemokines, proteolytic enzymes, adhesion molecules, cytokines, and stromal cells. Exp Hematol 2002; 30(9): 973-81.
[http://dx.doi.org/10.1016/S0301-472X(02)00883-4] [PMID: 12225788]

[16]    Belnoue E, Tougne C, Rochat AF, Lambert PH, Pinschewer DD, Siegrist CA. Homing and adhesion patterns determine the cellular composition of the bone marrow plasma cell niche. J Immunol 2012; 188(3): 1283-91.
[http://dx.doi.org/10.4049/jimmunol.1103169] [PMID: 22262758]

[17]    Lapidot T, Kollet O. The essential roles of the chemokine SDF-1 and its receptor CXCR4 in human stem cell homing and repopulation of transplanted immune-deficient NOD/SCID and NOD/SCID/B2m(null) mice. Leukemia 2002; 16(10): 1992-2003.
[http://dx.doi.org/10.1038/sj.leu.2402684] [PMID: 12357350]

[18]    Taylor EW. Cell motility. J Cell Sci Suppl 1986; 4: 89-102.
[http://dx.doi.org/10.1242/jcs.1986.Supplement_4.6] [PMID: 2943749]

[19]    Dean M, Fojo T, Bates S. Tumour stem cells and drug resistance. Nat Rev Cancer 2005; 5(4): 275-84.
[http://dx.doi.org/10.1038/nrc1590] [PMID: 15803154]

[20]    Jones CV, Copelan EA. Treatment of acute myeloid leukemia with hematopoietic stem cell transplantation. Future Oncol 2009; 5(4): 559-68.
[http://dx.doi.org/10.2217/fon.09.20] [PMID: 19450182]

[21]    Laurence AD. Location, movement and survival: the role of chemokines in haematopoiesis and malignancy. Br J Haematol 2006; 132(3): 255-67.
[http://dx.doi.org/10.1111/j.1365-2141.2005.05841.x] [PMID: 16409290]

[22]    Orkin SH. Diversification of haematopoietic stem cells to specific lineages. Nat Rev Genet 2000; 1(1): 57-64.
        [http://dx.doi.org/10.1038/35049577] [PMID: 11262875]

[23]    Zhai PF, Wang F, Su R, *et al.* The regulatory roles of microRNA-146b-5p and its target platelet-derived growth factor receptor α (PDGFRA) in erythropoiesis and megakaryocytopoiesis. J Biol Chem 2014; 289(33): 22600-13.
        [http://dx.doi.org/10.1074/jbc.M114.547380] [PMID: 24982425]

[24]    Magga J, Savchenko E, Malm T, *et al.* Production of monocytic cells from bone marrow stem cells: therapeutic usage in Alzheimers disease. J Cell Mol Med 2012; 16(5): 1060-73.
        [http://dx.doi.org/10.1111/j.1582-4934.2011.01390.x] [PMID: 21777378]

[25]    Li Y, Adomat H, Guns ET, *et al.* Identification of a hematopoietic cell dedifferentiation-inducing factor. J Cell Physiol 2016; 231(6): 1350-63.
        [http://dx.doi.org/10.1002/jcp.25239] [PMID: 26529564]

[26]    Schlessinger J. Cell signaling by receptor tyrosine kinases. Cell 2000; 103(2): 211-25.
        [http://dx.doi.org/10.1016/S0092-8674(00)00114-8] [PMID: 11057895]

[27]    Zon LI, Youssoufian H, Mather C, Lodish HF, Orkin SH. Activation of the erythropoietin receptor promoter by transcription factor GATA-1. Proc Natl Acad Sci USA 1991; 88(23): 10638-41.
        [http://dx.doi.org/10.1073/pnas.88.23.10638] [PMID: 1660143]

[28]    Timchenko N, Wilson DR, Taylor LR, *et al.* Autoregulation of the human C/EBP alpha gene by stimulation of upstream stimulatory factor binding. Mol Cell Biol 1995; 15(3): 1192-202.
        [http://dx.doi.org/10.1128/MCB.15.3.1192] [PMID: 7862113]

[29]    Osgood EE, Riddle MC, Mathew TJ. Aplastic anemia treated with daily transfusion and intravenous marrow: case report. Ann Intern Med 1939; 13: 357.
        [http://dx.doi.org/10.7326/0003-4819-13-2-357]

[30]    Gatti RA, Meuwissen HJ, Allen HD, Hong R, Good RA. Immunological reconstitution of sex-linked lymphopenic immunological deficiency. Lancet 1968; 2(7583): 1366-9.
        [http://dx.doi.org/10.1016/S0140-6736(68)92673-1] [PMID: 4177932]

[31]    Hong R, Cooper MD, Allan MJ, Kay HE, Meuwissen H, Good RA. Immunological restitution in lymphopenic immunological deficiency syndrome. Lancet 1968; 1(7541): 503-6.
        [http://dx.doi.org/10.1016/S0140-6736(68)91468-2] [PMID: 4171200]

[32]    Bach FH, Albertini RJ, Joo P, Anderson JL, Bortin MM. Bone-marrow transplantation in a patient with the Wiskott-Aldrich syndrome. Lancet 1968; 2(7583): 1364-6.
        [http://dx.doi.org/10.1016/S0140-6736(68)92672-X] [PMID: 4177931]

[33]    Liakapoulouand E, Marks DI. Introduction to haematopoietic stem cell transplantation. Treatment of Cancer. 5th ed., 2008.

[34]    Fefer A, Cheever MA, Greenberg PD, *et al.* Treatment of chronic granulocytic leukemia with chemoradiotherapy and transplantation of marrow from identical twins. N Engl J Med 1982; 306(2): 63-8.
        [http://dx.doi.org/10.1056/NEJM198201143060202] [PMID: 7031474]

[35] Thomas E, Storb R, Clift RA, *et al.* Bone-marrow transplantation (first of two parts). N Engl J Med 1975; 292(16): 832-43.
[http://dx.doi.org/10.1056/NEJM197504172921605] [PMID: 234595]

[36] Beatty PG, Mori M, Milford E. Impact of racial genetic polymorphism on the probability of finding an HLA-matched donor. Transplantation 1995; 60(8): 778-83.
[http://dx.doi.org/10.1097/00007890-199510270-00003] [PMID: 7482734]

[37] OReilly RJ. Allogenic bone marrow transplantation: current status and future directions. Blood 1983; 62(5): 941-64.
[PMID: 6354307]

[38] Velardi A, Ruggeri L, Mancusi A, *et al.* Clinical impact of natural killer cell reconstitution after allogeneic hematopoietic transplantation. Semin Immunopathol 2008; 30(4): 489-503.
[http://dx.doi.org/10.1007/s00281-008-0136-1] [PMID: 19002464]

[39] Ruggeri L, Mancusi A, Burchielli E, *et al.* Natural killer cell recognition of missing self and haploidentical hematopoietic transplantation. Semin Cancer Biol 2006; 16(5): 404-11.
[http://dx.doi.org/10.1016/j.semcancer.2006.07.007] [PMID: 16916611]

[40] Perruccio K, Topini F, Tosti A, *et al.* Optimizing a photoallodepletion protocol for adoptive immunotherapy after haploidentical SCT. Bone Marrow Transplant 2012; 47(9): 1196-200.
[http://dx.doi.org/10.1038/bmt.2011.237] [PMID: 22139067]

[41] Mo XD, Xu LP, Zhang XH, *et al.* Chronic GVHD induced GVL effect after unmanipulated haploidentical hematopoietic SCT for AML and myelodysplastic syndrome. Bone Marrow Transplant 2015; 50(1): 127-33.
[http://dx.doi.org/10.1038/bmt.2014.223] [PMID: 25387095]

[42] Wu T, Young JS, Johnston H, *et al.* Thymic damage, impaired negative selection, and development of chronic graft-versus-host disease caused by donor CD4+ and CD8+ T cells. J Immunol 2013; 191(1): 488-99.
[http://dx.doi.org/10.4049/jimmunol.1300657] [PMID: 23709681]

[43] Perruccio K, Topini F, Tosti A, *et al.* Differences in Aspergillus-specific immune recovery between T-cell-replete and T-cell-depleted hematopoietic transplants. Eur J Haematol 2015; 95(6): 551-7.
[http://dx.doi.org/10.1111/ejh.12531] [PMID: 25688598]

[44] Perruccio K, Topini F, Tosti A, *et al.* Photodynamic purging of alloreactive T cells for adoptive immunotherapy after haploidentical stem cell transplantation. Blood Cells Mol Dis 2008; 40(1): 76-83.
[http://dx.doi.org/10.1016/j.bcmd.2007.06.022] [PMID: 17977031]

[45] Fuchs E, ODonnell PV, Brunstein CG. Alternative transplant donor sources: is there any consensus? Curr Opin Oncol 2013; 25(2): 173-9.
[http://dx.doi.org/10.1097/CCO.0b013e32835d815f] [PMID: 23385861]

[46] Saber W, Opie S, Rizzo JD, Zhang MJ, Horowitz MM, Schriber J. Outcomes after matched unrelated donor versus identical sibling hematopoietic cell transplantation in adults with acute myelogenous leukemia. Blood 2012; 119(17): 3908-16.
[http://dx.doi.org/10.1182/blood-2011-09-381699] [PMID: 22327226]

[47] MacMillan ML, Wagner JE. Haematopoeitic cell transplantation for Fanconi anaemia - when and

how? Br J Haematol 2010; 149(1): 14-21.
[http://dx.doi.org/10.1111/j.1365-2141.2010.08078.x] [PMID: 20136826]

[48]    Kleen TO, Kadereit S, Fanning LR, *et al.* Recipient-specific tolerance after HLA-mismatched umbilical cord blood stem cell transplantation. Transplantation 2005; 80(9): 1316-22.
[http://dx.doi.org/10.1097/01.tp.0000188172.26531.6f] [PMID: 16314801]

[49]    Farnault L, Gertner-Dardenne J, Gondois-Rey F, *et al.* Full but impaired activation of innate immunity effectors and virus-specific T cells during CMV and EBV disease following cord blood transplantation. Bone Marrow Transplant 2015; 50(3): 459-62.
[http://dx.doi.org/10.1038/bmt.2014.275] [PMID: 25599160]

[50]    Sideri A, Neokleous N, Brunet De La Grange P, *et al.* An overview of the progress on double umbilical cord blood transplantation. Haematologica 2011; 96(8): 1213-20.
[http://dx.doi.org/10.3324/haematol.2010.038836] [PMID: 21546497]

[51]    Sullivan MJ. Banking on cord blood stem cells. Nat Rev Cancer 2008; 8(7): 555-63.
[http://dx.doi.org/10.1038/nrc2418] [PMID: 18548085]

[52]    Evans MD, Kelley J. US attitudes toward human embryonic stem cell research. Nat Biotechnol 2011; 29(6): 484-8.
[http://dx.doi.org/10.1038/nbt.1891] [PMID: 21654664]

[53]    Hansen JA, Petersdorf E, Martin PJ, Anasetti C. Hematopoietic stem cell transplants from unrelated donors. Immunol Rev 1997; 157: 141-51.
[http://dx.doi.org/10.1111/j.1600-065X.1997.tb00979.x] [PMID: 9255627]

[54]    Uharek L, Glass B, Gassmann W, *et al.* Engraftment of allogeneic bone marrow cells: experimental investigations on the role of cell dose, graft-*versus*-host reactive T cells and pretransplant immunosuppression. Transplant Proc 1992; 24(6): 3023-5.
[PMID: 1466042]

[55]    Reisner Y, Kapoor N, Kirkpatrick D, *et al.* Transplantation for acute leukaemia with HLA-A and B nonidentical parental marrow cells fractionated with soybean agglutinin and sheep red blood cells. Lancet 1981; 2(8242): 327-31.
[http://dx.doi.org/10.1016/S0140-6736(81)90647-4] [PMID: 6115110]

[56]    Prentice HG, Blacklock HA, Janossy G, *et al.* Depletion of T lymphocytes in donor marrow prevents significant graft-*versus*-host disease in matched allogeneic leukaemic marrow transplant recipients. Lancet 1984; 1(8375): 472-6.
[http://dx.doi.org/10.1016/S0140-6736(84)92848-4] [PMID: 6142207]

[57]    O'Reilly RJ, Collins NH, Brochstein J. Soybean lectin agglutination and E-rosette depletion for removal of T-cells from HLA-identical marrow grafts: results in 60 consecutive patients transplanted for hematological malignancy. In: Hagenbeek A, Löwenberg B, Eds. Minimal Residual Disease in Acute Leukemia. Martinus Nijhoff Publishers 1986; pp. 337-44.
[http://dx.doi.org/10.1007/978-94-009-4273-8_31]

[58]    Hale G, Cobbold S, Waldmann H. T cell depletion with CAMPATH-1 in allogeneic bone marrow transplantation. Transplantation 1988; 45(4): 753-9.
[http://dx.doi.org/10.1097/00007890-198804000-00018] [PMID: 3282358]

[59]    Marmont AM, Horowitz MM, Gale RP, *et al.* T-cell depletion of HLA-identical transplants in leukemia. Blood 1991; 78(8): 2120-30.
[PMID: 1912589]

[60]    Guinan EC, Boussiotis VA, Neuberg D, *et al.* Transplantation of anergic histoincompatible bone marrow allografts. N Engl J Med 1999; 340(22): 1704-14.
[http://dx.doi.org/10.1056/NEJM199906033402202] [PMID: 10352162]

[61]    Reisner Y, Ben-Bassat I, Douer D, Kaploon A, Schwartz E, Ramot B. Demonstration of clonable alloreactive host T cells in a primate model for bone marrow transplantation. Proc Natl Acad Sci USA 1986; 83(11): 4012-5.
[http://dx.doi.org/10.1073/pnas.83.11.4012] [PMID: 3520563]

[62]    Schwartz E, Lapidot T, Gozes D, Singer TS, Reisner Y. Abrogation of bone marrow allograft resistance in mice by increased total body irradiation correlates with eradication of host clonable T cells and alloreactive cytotoxic precursors. J Immunol 1987; 138(2): 460-5.
[PMID: 3098843]

[63]    Bachar-Lustig E, Rachamim N, Li HW, Lan F, Reisner Y. Megadose of T cell-depleted bone marrow overcomes MHC barriers in sublethally irradiated mice. Nat Med 1995; 1(12): 1268-73.
[http://dx.doi.org/10.1038/nm1295-1268] [PMID: 7489407]

[64]    Reisner Y, Martelli MF. Bone marrow transplantation across HLA barriers by increasing the number of transplanted cells. Immunol Today 1995; 16(9): 437-40.
[http://dx.doi.org/10.1016/0167-5699(95)80021-2] [PMID: 7546208]

[65]    Aversa F, Tabilio A, Terenzi A, *et al.* Successful engraftment of T-cell-depleted haploidentical three-loci incompatible transplants in leukemia patients by addition of recombinant human granulocyte colony-stimulating factor-mobilized peripheral blood progenitor cells to bone marrow inoculum. Blood 1994; 84(11): 3948-55.
[PMID: 7524753]

[66]    Perruccio K, Topini F, Tosti A, *et al.* Differences in *Aspergillus*-specific immune recovery between T-cell-replete and T-cell-depleted hematopoietic transplants. Eur J Haematol 2015; 95(6): 551-7.
[http://dx.doi.org/10.1111/ejh.12531] [PMID: 25688598]

[67]    Aversa F, Tabilio A, Velardi A, *et al.* Transplantation for high-risk acute leukemia with high doses of T cell-depleted hematopoietic stem cells from haploidentical three loci incompatible donors. N Engl J Med 1998; 339(17): 1186-93.
[http://dx.doi.org/10.1056/NEJM199810223391702] [PMID: 9780338]

[68]    Aversa F, Terenzi A, Tabilio A, *et al.* Full haplotype-mismatched hematopoietic stem-cell transplantation: a phase II study in patients with acute leukemia at high risk of relapse. J Clin Oncol 2005; 23(15): 3447-54.
[http://dx.doi.org/10.1200/JCO.2005.09.117] [PMID: 15753458]

[69]    Ruggeri L, Capanni M, Casucci M, *et al.* Role of natural killer cell alloreactivity in HLA-mismatched hematopoietic stem cell transplantation. Blood 1999; 94(1): 333-9.
[PMID: 10381530]

[70]    Ruggeri L, Capanni M, Urbani E, *et al.* Effectiveness of donor natural killer cell alloreactivity in

mismatched hematopoietic transplants. Science 2002; 295(5562): 2097-100.
[http://dx.doi.org/10.1126/science.1068440] [PMID: 11896281]

[71]    Mancusi A, Ruggeri L, Urbani E, *et al.* Haploidentical hematopoietic transplantation from KIR ligand-mismatched donors with activating KIRs reduces nonrelapse mortality. Blood 2015; 125(20): 3173-82.
[http://dx.doi.org/10.1182/blood-2014-09-599993] [PMID: 25769621]

[72]    Byrd JC, Dodge RK, Carroll A, *et al.* Patients with t(8;21)(q22;q22) and acute myeloid leukemia have superior failure-free and overall survival when repetitive cycles of high-dose cytarabine are administered. J Clin Oncol 1999; 17(12): 3767-75.
[PMID: 10577848]

[73]    Bloomfield CD, Lawrence D, Byrd JC, *et al.* Frequency of prolonged remission duration after high-dose cytarabine intensification in acute myeloid leukemia varies by cytogenetic subtype. Cancer Res 1998; 58(18): 4173-9.
[PMID: 9751631]

[74]    Adès L, Sanz MA, Chevret S, *et al.* Treatment of newly diagnosed acute promyelocytic leukemia (APL): a comparison of French-Belgian-Swiss and PETHEMA results. Blood 2008; 111(3): 1078-84.
[http://dx.doi.org/10.1182/blood-2007-07-099978] [PMID: 17975017]

[75]    Armand P, Kim HT, DeAngelo DJ, *et al.* Impact of cytogenetics on outcome of *de novo* and therapy-related AML and MDS after allogeneic transplantation. Biol Blood Marrow Transplant 2007; 13(6): 655-64.
[http://dx.doi.org/10.1016/j.bbmt.2007.01.079] [PMID: 17531775]

[76]    Chang C, Storer BE, Scott BL, *et al.* Hematopoietic cell transplantation in patients with myelodysplastic syndrome or acute myeloid leukemia arising from myelodysplastic syndrome: similar outcomes in patients with *de novo* disease and disease following prior therapy or antecedent hematologic disorders. Blood 2007; 110(4): 1379-87.
[http://dx.doi.org/10.1182/blood-2007-02-076307] [PMID: 17488876]

[77]    Smith SM, Le Beau MM, Huo D, *et al.* Clinical-cytogenetic associations in 306 patients with therapy-related myelodysplasia and myeloid leukemia: the University of Chicago series. Blood 2003; 102(1): 43-52.
[http://dx.doi.org/10.1182/blood-2002-11-3343] [PMID: 12623843]

[78]    Slovak ML, Kopecky KJ, Cassileth PA, *et al.* Karyotypic analysis predicts outcome of permission and postremission therapy in adult acute myeloid leukemia: a Southwest Oncology Group/Eastern Cooperative Oncology Group Study. Blood 2000; 96(13): 4075-83.
[PMID: 11110676]

[79]    Yanada M, Matsuo K, Emi N, Naoe T. Efficacy of allogeneic hematopoietic stem cell transplantation depends on cytogenetic risk for acute myeloid leukemia in first disease remission: a metaanalysis. Cancer 2005; 103(8): 1652-8.
[http://dx.doi.org/10.1002/cncr.20945] [PMID: 15742336]

[80]    Krauter J, Heil G, Hoelzer D. Role of consolidation therapy in the treatment of patients up to 60 years with high risk AML. Blood 2005; 106: 172a.

[81]    Schlenk RF, Döhner K, Krauter J, *et al.* Mutations and treatment outcome in cytogenetically normal acute myeloid leukemia. N Engl J Med 2008; 358(18): 1909-18.

[http://dx.doi.org/10.1056/NEJMoa074306] [PMID: 18450602]

[82]  Archimbaud E, Thomas X, Michallet M, *et al.* Prospective genetically randomized comparison between intensive postinduction chemotherapy and bone marrow transplantation in adults with newly diagnosed acute myeloid leukemia. J Clin Oncol 1994; 12(2): 262-7.
[PMID: 8113835]

[83]  Sekeres MA, Stone RM, Zahrieh D, *et al.* Decision-making and quality of life in older adults with acute myeloid leukemia or advanced myelodysplastic syndrome. Leukemia 2004; 18(4): 809-16.
[http://dx.doi.org/10.1038/sj.leu.2403289] [PMID: 14762444]

[84]  Sorror ML, Maris MB, Storb R, *et al.* Hematopoietic cell transplantation (HCT)-specific comorbidity index: a new tool for risk assessment before allogeneic HCT. Blood 2005; 106(8): 2912-9.
[http://dx.doi.org/10.1182/blood-2005-05-2004] [PMID: 15994282]

[85]  Sorror ML, Giralt S, Sandmaier BM, *et al.* Hematopoietic cell transplantation specific comorbidity index as an outcome predictor for patients with acute myeloid leukemia in first remission: combined FHCRC and MDACC experiences. Blood 2007; 110(13): 4606-13.
[http://dx.doi.org/10.1182/blood-2007-06-096966] [PMID: 17873123]

[86]  Giles FJ, Borthakur G, Ravandi F, *et al.* The haematopoietic cell transplantation comorbidity index score is predictive of early death and survival in patients over 60 years of age receiving induction therapy for acute myeloid leukaemia. Br J Haematol 2007; 136(4): 624-7.
[http://dx.doi.org/10.1111/j.1365-2141.2006.06476.x] [PMID: 17223919]

[87]  Zittoun RA, Mandelli F, Willemze R, *et al.* Autologous or allogeneic bone marrow transplantation compared with intensive chemotherapy in acute myelogenous leukemia. European Organization for Research and Treatment of Cancer (EORTC) and the Gruppo Italiano Malattie Ematologiche Maligne dellAdulto (GIMEMA) Leukemia Cooperative Groups. N Engl J Med 1995; 332(4): 217-23.
[http://dx.doi.org/10.1056/NEJM199501263320403] [PMID: 7808487]

[88]  Appelbaum FR, Pearce SF. Hematopoietic cell transplantation in first complete remission *versus* early relapse. Best Pract Res Clin Haematol 2006; 19(2): 333-9.
[http://dx.doi.org/10.1016/j.beha.2005.12.001] [PMID: 16516131]

[89]  Jones CV, Copelan EA. Treatment of acute myeloid leukemia with hematopoietic stem cell transplantation. Future Oncol 2009; 5(4): 559-68.
[http://dx.doi.org/10.2217/fon.09.20] [PMID: 19450182]

[90]  Cornelissen JJ, van Putten WL, Verdonck LF, *et al.* Results of a HOVON/SAKK donor *versus* no-donor analysis of myeloablative HLA-identical sibling stem cell transplantation in first remission acute myeloid leukemia in young and middle-aged adults: benefits for whom? Blood 2007; 109(9): 3658-66.
[http://dx.doi.org/10.1182/blood-2006-06-025627] [PMID: 17213292]

[91]  Cutler CS, Lee SJ, Greenberg P, *et al.* A decision analysis of allogeneic bone marrow transplantation for the myelodysplastic syndromes: delayed transplantation for low-risk myelodysplasia is associated with improved outcome. Blood 2004; 104(2): 579-85.
[http://dx.doi.org/10.1182/blood-2004-01-0338] [PMID: 15039286]

[92]  MEI Pharma Reports Updated Results from Phase II Study of Pracinostat and Azacitidine in Elderly Patients with newly Diagnosed Acute Myeloid Leukemia 2015.

[93]    Malcovati L, Germing U, Kuendgen A, *et al.* Time-dependent prognostic scoring system for predicting survival and leukemic evolution in myelodysplastic syndromes. J Clin Oncol 2007; 25(23): 3503-10. [http://dx.doi.org/10.1200/JCO.2006.08.5696] [PMID: 17687155]

[94]    Navarro I, Ruiz MA, Cabello A, *et al.* Classification and scoring systems in myelodysplastic syndromes: a retrospective analysis of 311 patients. Leuk Res 2006; 30(8): 971-7. [http://dx.doi.org/10.1016/j.leukres.2005.11.015] [PMID: 16423393]

[95]    Park MJ, Kim HJ, Kim SH, *et al.* Is International Prognostic Scoring System (IPSS) still standard in predicting prognosis in patients with myelodysplastic syndrome? External validation of the WHO Classification-Based Prognostic Scoring System (WPSS) and comparison with IPSS. Eur J Haematol 2008; 81(5): 364-73. [PMID: 18637029]

[96]    Alessandrino EP, Della Porta MG, Bacigalupo A, *et al.* Gruppo Italiano Trapianto di Midollo Osseo (GITMO). WHO classification and WPSS predict posttransplantation outcome in patients with myelodysplastic syndrome: a study from the Gruppo Italiano Trapianto di Midollo Osseo (GITMO). Blood 2008; 112(3): 895-902. [http://dx.doi.org/10.1182/blood-2008-03-143735] [PMID: 18497321]

[97]    Ma L, Hao S, Diong C, *et al.* WPSS is a strong prognostic indicator for clinical outcome of allogeneic transplant for myelodysplastic syndrome in Southeast Asian patients. Ann Hematol 2015; 94(5): 761-9. [http://dx.doi.org/10.1007/s00277-014-2275-x] [PMID: 25519475]

[98]    Sebban C, Lepage E, Vernant JP, *et al.* Allogeneic bone marrow transplantation in adult acute lymphoblastic leukemia in first complete remission: a comparative study. J Clin Oncol 1994; 12(12): 2580-7. [PMID: 7989932]

[99]    Thomas X, Boiron JM, Huguet F, *et al.* Outcome of treatment in adults with acute lymphoblastic leukemia: analysis of the LALA-94 trial. J Clin Oncol 2004; 22(20): 4075-86. [http://dx.doi.org/10.1200/JCO.2004.10.050] [PMID: 15353542]

[100]   Chen Y, Wang H, Kantarjian H, Cortes J. Trends in chronic myeloid leukemia incidence and survival in the United States from 1975 to 2009. Leuk Lymphoma 2013; 54(7): 1411-7. [http://dx.doi.org/10.3109/10428194.2012.745525] [PMID: 23121646]

[101]   Jabbour E, Kantarjian H, OBrien S, *et al.* Predictive factors for outcome and response in patients treated with second-generation tyrosine kinase inhibitors for chronic myeloid leukemia in chronic phase after imatinib failure. Blood 2011; 117(6): 1822-7. [http://dx.doi.org/10.1182/blood-2010-07-293977] [PMID: 21030554]

[102]   A new co-operative group: the European Myeloma Network Trialist Group. Leuk Lymphoma 2009; 50(10): 1581-8. [http://dx.doi.org/10.1080/10428190903191874] [PMID: 19757308]

[103]   Seyfizadeh N, Seyfizadeh N, Hasenkamp J, Huerta-Yepez S. A molecular perspective on rituximab: A monoclonal antibody for B cell non Hodgkin lymphoma and other affections. Crit Rev Oncol Hematol 2016; 97: 275-90. [http://dx.doi.org/10.1016/j.critrevonc.2015.09.001] [PMID: 26443686]

[104]  Chen R, Palmer JM, Martin P, *et al.* Results of a multicenter phase II trial of Brentuximab Vedotin as second-line therapy before autologous transplantation in relapsed/refractory Hodgkin-lymphoma. Biol Blood Marrow Transplant 2015; 21(12): 2136-40.
[http://dx.doi.org/10.1016/j.bbmt.2015.07.018] [PMID: 26211987]

[105]  Tanimoto TE, Kusumi E, Hamaki T, *et al.* High complete response rate after allogeneic hematopoietic stem cell transplantation with reduced-intensity conditioning regimens in advanced malignant lymphoma. Bone Marrow Transplant 2003; 32(2): 131-7.
[http://dx.doi.org/10.1038/sj.bmt.1704118] [PMID: 12838276]

[106]  Seyfarth B, Josting A, Dreyling M, Schmitz N. Relapse in common lymphoma subtypes: salvage treatment options for follicular lymphoma, diffuse large cell lymphoma and Hodgkin disease. Br J Haematol 2006; 133(1): 3-18.
[http://dx.doi.org/10.1111/j.1365-2141.2006.05975.x] [PMID: 16512824]

[107]  Toriyama K. Ethnogeographical coincidence of endemic Kaposi's sarcoma and African Burkitt's lymphoma in western Kenya. Japanese Journal of Tropical Medicine and Hygiene 1988.
[http://dx.doi.org/10.2149/tmh1973.16.85]

[108]  Lanzkowsky P. Bone Marrow Failure. Manual of Pediatric Hematology Oncology. 5$^{th}$ ed. 2011; pp. 123-67.
[http://dx.doi.org/10.1016/B978-0-12-375154-6.00006-9]

[109]  Bacigalupo A, Passweg J. Diagnosis and treatment of acquired aplastic anemia. Hematol Oncol Clin North Am 2009; 23(2): 159-70.
[http://dx.doi.org/10.1016/j.hoc.2009.01.005] [PMID: 19327577]

[110]  Bacigalupo A, Socié G, Hamladji RM, *et al.* Current outcome of HLA identical sibling *versus* unrelated donor transplants in severe aplastic anemia: an EBMT analysis. Haematologica 2015; 100(5): 696-702.
[http://dx.doi.org/10.3324/haematol.2014.115345] [PMID: 25616576]

[111]  Bacigalupo A, Socie G, Lanino E, *et al.* Fludarabine, cyclophosphamide, antithymocyte globulin, with or without low dose total body irradiation, for alternative donor transplants, in acquired severe aplastic anemia: a retrospective study from the EBMT-SAA Working Party. Haematologica 2010; 95(6): 976-82.
[http://dx.doi.org/10.3324/haematol.2009.018267] [PMID: 20494932]

[112]  Bacigalupo A. Aplastic anemia: pathogenesis and treatment. Hematology 2007; 23-8.
[http://dx.doi.org/10.1182/asheducation-2007.1.23] [PMID: 18024605]

[113]  Davies JK, Guinan EC. An update on the management of severe idiopathic aplastic anaemia in children. Br J Haematol 2007; 136(4): 549-64.
[http://dx.doi.org/10.1111/j.1365-2141.2006.06461.x] [PMID: 17214739]

[114]  Dufour C, Pillon M, Passweg J, *et al.* Outcome of aplastic anemia in adolescence: a survey of the severe aplastic anemia working party of the European group for blood and marrow transplantation. Haematologica 2014; 99(10): 1574-81.
[http://dx.doi.org/10.3324/haematol.2014.106096] [PMID: 25085353]

[115]  Locatelli F, Bruno B, Zecca M, *et al.* Cyclosporin A and short-term methotrexate *versus* cyclosporin A

as graft *versus* host disease prophylaxis in patients with severe aplastic anemia given allogeneic bone marrow transplantation from an HLA-identical sibling: results of a GITMO/EBMT randomized trial. Blood 2000; 96(5): 1690-7.
[PMID: 10961865]

[116]   Marsh JC, Kulasekararaj AG. Management of the refractory aplastic anemia: what are the options?. Hematology Am Soc Hematol Educ Program 2013; 87-94.
[http://dx.doi.org/10.1182/asheducation-2013.1.87]

[117]   Peffault de Latour R, Peters C, Gibson B, *et al.* Recommendations on hematopoietic stem cell transplantation for inherited bone marrow failure syndromes. Bone Marrow Transplant 2015; 50(9): 1168-72.
[http://dx.doi.org/10.1038/bmt.2015.117] [PMID: 26052913]

[118]   Peffault de Latour R, Porcher R, Dalle JH, *et al.* FA Committee of the Severe Aplastic Anemia Working Party; Pediatric Working Party of the European Group for Blood and Marrow Transplantation. Allogeneic hematopoietic stem cell transplantation in Fanconi anemia: the European Group for Blood and Marrow Transplantation experience. Blood 2013; 122(26): 4279-86.
[http://dx.doi.org/10.1182/blood-2013-01-479733] [PMID: 24144640]

[119]   MacMillan ML, Wagner JE. Haematopoeitic cell transplantation for Fanconi anaemia - when and how? Br J Haematol 2010; 149(1): 14-21.
[http://dx.doi.org/10.1111/j.1365-2141.2010.08078.x] [PMID: 20136826]

[120]   Ahmed M, Dokal I. Understanding aplastic anaemia/bone-marrow failure syndromes. Paediatr Child Health 2009; 19(8): 351-7.

[121]   Lekstorm-Himes JA, Gallin JI. Immunodeficiency diseases caused by defect in phagocytes. N Engl Med 2000; 343(23): 1703-14.
[http://dx.doi.org/10.1056/NEJM200012073432307]

[122]   Wanachiwanawin W, Siripanyaphinyo U, Piyawattanasakul N, Kinoshita T. A cohort study of the nature of paroxysmal nocturnal hemoglobinuria clones and PIG-A mutations in patients with aplastic anemia. Eur J Haematol 2006; 76(6): 502-9.
[http://dx.doi.org/10.1111/j.0902-4441.2005.t01-1-EJH2467.x] [PMID: 16529603]

[123]   Peffault de Latour R, Schrezenmeier H, Bacigalupo A, *et al.* Allogeneic stem cell transplantation in paroxysmal nocturnal hemoglobinuria. Haematologica 2012; 97(11): 1666-73.
[http://dx.doi.org/10.3324/haematol.2012.062828] [PMID: 22689687]

[124]   Renella R, Wood WG. The congenital dyserythropoietic anemias. Hematol Oncol Clin North Am 2009; 23(2): 283-306.
[http://dx.doi.org/10.1016/j.hoc.2009.01.010] [PMID: 19327584]

[125]   Lucarelli G, Isgrò A, Sodani P, Gaziev J. Hematopoietic stem cell transplantation in thalassemia and sickle cell anemia. Cold Spring Harb Perspect Med 2012; 2(5): a011825.
[http://dx.doi.org/10.1101/cshperspect.a011825] [PMID: 22553502]

[126]   Sodani P, Isgrò A, Gaziev J, *et al.* T cell-depleted hla-haploidentical stem cell transplantation in thalassemia young patients. Pediatr Rep 2011; 3 (Suppl. 2): e13.
[http://dx.doi.org/10.4081/pr.2011.s2.e12] [PMID: 22053275]

[127] Hsieh MM, Kang EM, Fitzhugh CD, *et al.* Allogeneic hematopoietic stem-cell transplantation for sickle cell disease. N Engl J Med 2009; 361(24): 2309-17.
[http://dx.doi.org/10.1056/NEJMoa0904971] [PMID: 20007560]

[128] Sureda A, Bader P, Cesaro S, *et al.* Indications for allo- and auto-SCT for haematological diseases, solid tumours and immune disorders: current practice in Europe, 2015. Bone Marrow Transplant 2015; 50(8): 1037-56.
[http://dx.doi.org/10.1038/bmt.2015.6] [PMID: 25798672]

[129] Kletzel M, Katzenstein HM, Haut PR, *et al.* Treatment of high-risk neuroblastoma with triple-tandem high-dose therapy and stem-cell rescue: results of the Chicago Pilot II Study. J Clin Oncol 2002; 20(9): 2284-92.
[http://dx.doi.org/10.1200/JCO.2002.06.060] [PMID: 11980999]

[130] Hendershot E. Solid Tumors. Pediatric Oncology Nursing, in pediatric oncology 2005; 59-127.
[http://dx.doi.org/10.1007/3-540-26784-0_2]

[131] Kreissman SG, Rackoff W, Lee M, Breitfeld PP. High dose cyclophosphamide with carboplatin: a tolerable regimen suitable for dose intensification in children with solid tumors. J Pediatr Hematol Oncol 1997; 19(4): 309-12.
[http://dx.doi.org/10.1097/00043426-199707000-00008] [PMID: 9256829]

[132] Paul A. Meyers Systemic therapy for osteosarcoma and Ewing sarcoma. Asco Educational Book 2015.

[133] Lashkari A, Chow WA, Valdes F, *et al.* Tandem high-dose chemotherapy followed by autologous transplantation in patients with locally advanced or metastatic sarcoma. Anticancer Res 2009; 29(8): 3281-8.
[PMID: 19661346]

[134] Peinemann F, Kröger N, Bartel C, *et al.* High-dose chemotherapy followed by autologous stem cell transplantation for metastatic rhabdomyosarcomaa systematic review. PLoS One 2011; 6(2): e17127.
[http://dx.doi.org/10.1371/journal.pone.0017127] [PMID: 21373200]

[135] Koscielniak E, Klingebiel TH, Peters C, *et al.* Do patients with metastatic and recurrent rhabdomyosarcoma benefit from high-dose therapy with hematopoietic rescue? Report of the German/Austrian Pediatric Bone Marrow Transplantation Group. Bone Marrow Transplant 1997; 19(3): 227-31.
[http://dx.doi.org/10.1038/sj.bmt.1700628] [PMID: 9028550]

[136] Hamilton BK, Rybicki L, Abounader D, *et al.* Long-term survival after high-dose chemotherapy with autologous hematopoietic cell transplantation in metastatic breast cancer. Hematol Oncol Stem Cell Ther 2015; 8(3): 115-24.
[http://dx.doi.org/10.1016/j.hemonc.2015.06.005] [PMID: 26183670]

[137] Bregni M, Ciceri F, Peccatori J. Allogeneic stem cell transplantation for metastatic renal cell cancer (RCC). J Cancer 2011; 2: 347-9.
[http://dx.doi.org/10.7150/jca.2.347] [PMID: 21716855]

[138] Williams SD, Loehrer PJ, Nichols CR, Einhorn LN. Chemotherapy of male and female germ cell tumors. Semin Oncol 1992; 19(2) (Suppl. 5): 19-23.
[PMID: 1384139]

[139] Spriggs D. Optimal sequencing in the treatment of recurrent ovarian cancer. Gynecol Oncol 2003; 90(3 Pt 2): S39-44.
[http://dx.doi.org/10.1016/S0090-8258(03)00471-2] [PMID: 13129495]

[140] Mandanas RA, Saez RA, Epstein RB, Confer DL, Selby GB. Long-term results of autologous marrow transplantation for relapsed or refractory male or female germ cell tumors. Bone Marrow Transplant 1998; 21(6): 569-76.
[http://dx.doi.org/10.1038/sj.bmt.1701132] [PMID: 9543060]

[141] Loehrer PJ Sr, Einhorn LH, Williams SD. VP-16 plus ifosfamide plus cisplatin as salvage therapy in refractory germ cell cancer. J Clin Oncol 1986; 4(4): 528-36.
[PMID: 3633952]

[142] Einhorn LH. Salvage therapy for germ cell tumors. Semin Oncol 1994; 21(4) (Suppl. 7): 47-51.
[PMID: 8091241]

[143] Chen AR. High-dose therapy with stem cell rescue for pediatric solid tumors: rationale and results. Pediatr Transplant 1999; 3(1) (Suppl. 1): 78-86.
[http://dx.doi.org/10.1034/j.1399-3046.1999.00053.x] [PMID: 10587976]

[144] Leavey PJ, Collier AB. Ewing sarcoma: prognostic criteria, outcomes and future treatment. Expert Rev Anticancer Ther 2008; 8(4): 617-24.
[http://dx.doi.org/10.1586/14737140.8.4.617] [PMID: 18402528]

[145] Leavey PJ, Mascarenhas L, Marina N, *et al.* Prognostic factors for patients with Ewing sarcoma (EWS) at first recurrence following multi-modality therapy: A report from the Childrens Oncology Group. Pediatr Blood Cancer 2008; 51(3): 334-8.
[http://dx.doi.org/10.1002/pbc.21618] [PMID: 18506764]

[146] Miser JS, Goldsby RE, Chen Z, *et al.* Treatment of metastatic Ewing sarcoma/primitive neuroectodermal tumor of bone: evaluation of increasing the dose intensity of chemotherapy report from the Childrens Oncology Group. Pediatr Blood Cancer 2007; 49(7): 894-900.
[http://dx.doi.org/10.1002/pbc.21233] [PMID: 17584910]

[147] Meyers PA, Krailo MD, Ladanyi M, *et al.* High-dose melphalan, etoposide, total-body irradiation, and autologous stem-cell reconstitution as consolidation therapy for high-risk Ewings sarcoma does not improve prognosis. J Clin Oncol 2001; 19(11): 2812-20.
[PMID: 11387352]

[148] Burdach S, Meyer-Bahlburg A, Laws HJ, *et al.* High-dose therapy for patients with primary multifocal and early relapsed Ewings tumors: results of two consecutive regimens assessing the role of total-body irradiation. J Clin Oncol 2003; 21(16): 3072-8.
[http://dx.doi.org/10.1200/JCO.2003.12.039] [PMID: 12915596]

[149] Burdach S, van Kaick B, Laws HJ, *et al.* Allogeneic and autologous stem-cell transplantation in advanced Ewing tumors. An update after long-term follow-up from two centers of the European Intergroup study EICESS. Stem-Cell Transplant Programs at Düsseldorf University Medical Center, Germany and St. Anna Kinderspital, Vienna, Austria. Ann Oncol 2000; 11(11): 1451-62.
[http://dx.doi.org/10.1023/A:1026539908115] [PMID: 11142486]

[150] Oberlin O, Rey A, Desfachelles AS, *et al.* Impact of high-dose busulfan plus melphalan as

consolidation in metastatic Ewing tumors: a study by the Société Française des Cancers de lEnfant. J Clin Oncol 2006; 24(24): 3997-4002.
[http://dx.doi.org/10.1200/JCO.2006.05.7059] [PMID: 16921053]

[151] Luksch R, Tienghi A, Hall KS, *et al.* Primary metastatic Ewings family tumors: results of the Italian Sarcoma Group and Scandinavian Sarcoma Group ISG/SSG IV Study including myeloablative chemotherapy and total-lung irradiation. Ann Oncol 2012; 23(11): 2970-6.
[http://dx.doi.org/10.1093/annonc/mds117] [PMID: 22771824]

[152] Bacci G, Ferrari S, Longhi A, *et al.* Therapy and survival after recurrence of Ewings tumors: the Rizzoli experience in 195 patients treated with adjuvant and neoadjuvant chemotherapy from 1979 to 1997. Ann Oncol 2003; 14(11): 1654-9.
[http://dx.doi.org/10.1093/annonc/mdg457] [PMID: 14581274]

[153] Barker LM, Pendergrass TW, Sanders JE, Hawkins DS. Survival after recurrence of Ewings sarcoma family of tumors. J Clin Oncol 2005; 23(19): 4354-62.
[http://dx.doi.org/10.1200/JCO.2005.05.105] [PMID: 15781881]

[154] Ladenstein R, Philip T, Gardner H. Autologous stem cell transplantation for solid tumors in children. Curr Opin Pediatr 1997; 9(1): 55-69.
[http://dx.doi.org/10.1097/00008480-199702000-00013] [PMID: 9088757]

[155] Ferrari S, Sundby Hall K, Luksch R, *et al.* Nonmetastatic Ewing family tumors: high-dose chemotherapy with stem cell rescue in poor responder patients. Results of the Italian Sarcoma Group/Scandinavian Sarcoma Group III protocol. Ann Oncol 2011; 22(5): 1221-7.
[http://dx.doi.org/10.1093/annonc/mdq573] [PMID: 21059639]

[156] Maris JM, Hogarty MD, Bagatell R, Cohn SL. Neuroblastoma. Lancet 2007; 369(9579): 2106-20.
[http://dx.doi.org/10.1016/S0140-6736(07)60983-0] [PMID: 17586306]

[157] Seif AE, Naranjo A, Baker DL, *et al.* A pilot study of tandem high-dose chemotherapy with stem cell rescue as consolidation for high-risk neuroblastoma: Childrens Oncology Group study ANBL00P1. Bone Marrow Transplant 2013; 48(7): 947-52.
[http://dx.doi.org/10.1038/bmt.2012.276] [PMID: 23334272]

[158] Yu AL, Gilman AL, Ozkaynak MF, *et al.* Anti-GD2 antibody with GM-CSF, interleukin-2, and isotretinoin for neuroblastoma. N Engl J Med 2010; 363(14): 1324-34.
[http://dx.doi.org/10.1056/NEJMoa0911123] [PMID: 20879881]

[159] Hale GA, Arora M, Ahn KW, *et al.* Allogeneic hematopoietic cell transplantation for neuroblastoma: the CIBMTR experience. Bone Marrow Transplant 2013; 48(8): 1056-64.
[http://dx.doi.org/10.1038/bmt.2012.284] [PMID: 23419433]

[160] Gains J, Mandeville H, Cork N, Brock P, Gaze M. Ten challenges in the management of neuroblastoma. Future Oncol 2012; 8(7): 839-58.
[http://dx.doi.org/10.2217/fon.12.70] [PMID: 22830404]

[161] Boelens JJ, Prasad VK, Tolar J, Wynn RF, Peters C. Current international perspectives on hematopoietic stem cell transplantation for inherited metabolic disorders. Pediatr Clin North Am 2010; 57(1): 123-45.
[http://dx.doi.org/10.1016/j.pcl.2009.11.004] [PMID: 20307715]

[162] Cavazzana-Calvo M, Hacein-Bey S, de Saint Basile G, *et al.* Gene therapy of human severe combined immunodeficiency (SCID)-X1 disease. Science 2000; 288(5466): 669-72.
[http://dx.doi.org/10.1126/science.288.5466.669] [PMID: 10784449]

[163] McCormack MP, Rabbitts TH. Activation of the T-cell oncogene LMO2 after gene therapy for X-linked severe combined immunodeficiency. N Engl J Med 2004; 350(9): 913-22.
[http://dx.doi.org/10.1056/NEJMra032207] [PMID: 14985489]

[164] Ozsahin H, Cavazzana-Calvo M, Notarangelo LD, *et al.* Long-term outcome following hematopoietic stem-cell transplantation in Wiskott-Aldrich syndrome: collaborative study of the European Society for Immunodeficiencies and European Group for Blood and Marrow Transplantation. Blood 2008; 111(1): 439-45.
[http://dx.doi.org/10.1182/blood-2007-03-076679] [PMID: 17901250]

[165] Neven B, Leroy S, Decaluwe H, *et al.* Long-term outcome after hematopoietic stem cell transplantation of a single-center cohort of 90 patients with severe combined immunodeficiency. Blood 2009; 113(17): 4114-24.
[http://dx.doi.org/10.1182/blood-2008-09-177923] [PMID: 19168787]

[166] Fernandes JF, Rocha V, Labopin M, *et al.* Transplanting patients with severe combined immune deficiencies (SCID): mismatched related stem cells or unrelated cord blood? Blood 2012; 119(12): 2949-55.
[http://dx.doi.org/10.1182/blood-2011-06-363572] [PMID: 22308292]

[167] Pai SY, Logan BR, Griffith LM, Buckley RH, Parrott RE, Dvorak CC. Transplantation outcomes for severe combined immunodeficiency, 20002009. N Engl J Med 2014; 371(5): 434-46.
[http://dx.doi.org/10.1056/NEJMoa1401177] [PMID: 25075835]

[168] Witkowska M, Smolewska E, Smolewski P. Hematopoietic stem cell transplantation in children with autoimmune connective tissue diseases. Arch Immunol Ther Exp (Warsz) 2014; 62(4): 319-27.
[http://dx.doi.org/10.1007/s00005-014-0279-9] [PMID: 24604327]

[169] Boelens JJ, Orchard PJ, Wynn RF. Transplantation in inborn errors of metabolism: current considerations and future perspectives. Br J Haematol 2014; 167(3): 293-303.
[http://dx.doi.org/10.1111/bjh.13059] [PMID: 25074667]

[170] DAddio F, Valderrama Vasquez A, Ben Nasr M, *et al.* Autologous nonmyeloablative hematopoietic stem cell transplantation in new-onset type 1 diabetes: a multicenter analysis. Diabetes 2014; 63(9): 3041-6.
[http://dx.doi.org/10.2337/db14-0295] [PMID: 24947362]

[171] Skyler JS. Immune intervention for type 1 diabetes, 20132014. Diabetes Technol Ther 2015; 17(1) (Suppl. 1): S80-7.
[http://dx.doi.org/10.1089/dia.2015.1510] [PMID: 25679434]

[172] Perruccio K, Tosti A, Burchielli E, *et al.* Transferring functional immune responses to pathogens after haploidentical hematopoietic transplantation. Blood 2005; 106(13): 4397-406.
[http://dx.doi.org/10.1182/blood-2005-05-1775] [PMID: 16123217]

[173] Bertaina A, Merli P, Rutella S, *et al.* HLA-haploidentical stem cell transplantation after removal of αβ+ T and B cells in children with nonmalignant disorders. Blood 2014; 124(5): 822-6.

[http://dx.doi.org/10.1182/blood-2014-03-563817] [PMID: 24869942]

[174] Airoldi I, Bertaina A, Prigione I, *et al.* γδ T-cell reconstitution after HLA-haploidentical hematopoietic transplantation depleted of TCR-αβ+/CD19+ lymphocytes. Blood 2015; 125(15): 2349-58.
[http://dx.doi.org/10.1182/blood-2014-09-599423] [PMID: 25612623]

[175] Di Ianni M, Falzetti F, Carotti A, *et al.* Tregs prevent GVHD and promote immune reconstitution in HLA-haploidentical transplantation. Blood 2011; 117(14): 3921-8.
[http://dx.doi.org/10.1182/blood-2010-10-311894] [PMID: 21292771]

[176] Martelli MF, Di Ianni M, Ruggeri L, *et al.* HLA-haploidentical transplantation with regulatory and conventional T-cell adoptive immunotherapy prevents acute leukemia relapse. Blood 2014; 124(4): 638-44.
[http://dx.doi.org/10.1182/blood-2014-03-564401] [PMID: 24923299]

[177] Huang XJ, Han W, Xu LP, *et al.* A novel approach to human leukocyte antigen-mismatched transplantation in patients with malignant hematological disease. Chin Med J (Engl) 2004; 117(12): 1778-85.
[PMID: 15603704]

[178] Peccatori J, Forcina A, Clerici D, *et al.* Sirolimus-based graft-*versus*-host disease prophylaxis promotes the *in vivo* expansion of regulatory T cells and permits peripheral blood stem cell transplantation from haploidentical donors. Leukemia 2015; 29(2): 396-405.
[http://dx.doi.org/10.1038/leu.2014.180] [PMID: 24897508]

[179] Luznik L, ODonnell PV, Symons HJ, *et al.* HLA-haploidentical bone marrow transplantation for hematologic malignancies using nonmyeloablative conditioning and high-dose, posttransplantation cyclophosphamide. Biol Blood Marrow Transplant 2008; 14(6): 641-50.
[http://dx.doi.org/10.1016/j.bbmt.2008.03.005] [PMID: 18489989]

[180] Munchel A, Kesserwan C, Symons HJ, *et al.* Nonmyeloablative, HLA-haploidentical bone marrow transplantation with high dose, post-transplantation cyclophosphamide. Pediatr Rep 2011; 3 (Suppl. 2): e15.
[http://dx.doi.org/10.4081/pr.2011.s2.e15] [PMID: 22053277]

# CHAPTER 6

# Human Induced Pluripotent Stem Cells-Based Strategies: New Frontiers for Personalized Medicine

**Rosa Valentina Talarico, Giuseppe Novelli, Federica Sangiuolo** and **Paola Spitalieri**[*]

*Department of Biomedicine and Prevention, Tor Vergata University of Rome, Italy*

**Abstract:** Recent advanced protocols on cell reprogramming for the generation of human induced pluripotent stem cells (hiPSCs) has improved the comprehension of the pathogenic mechanisms and the development of new drugs. In fact, disease-specific pluripotent stem cells offer an ideal platform for both cell and gene therapy protocol applications and represent a good possibility for new and personalized pharmacological treatments. Without any doubt, the most innovative therapies are those which provide a site specific gene correction, and are suitable to those diseases for which a drug's therapy is not available. In the last decade have emerged ZFNs, TALENs, and the CRISPR/Cas9 system, tools for genome engineering, consisting of a sequence-specific DNA-binding domain and a non-specific DNA cleavage domain, that allow to correct mutated genes *in vitro*.

In this chapter, we focus on hiPSCs as a target cells for gene manipulation: new strategies as Zinc-finger nucleases, TALENs and CRISPR/ Cas9 have been developed to maximize the efficiency of genome editing protocols on human reprogrammed cells. Indeed, humanized iPSCs-based disease model systems exploit an individualized cell-based platform that has unlimited growth potential for novel regenerative strategies and clinical therapeutics, along with companion diagnostics, to predict and prognosticate the molecular basis of various human diseases.

[*] **Corresponding author Paola Spitalieri:** Department of Biomedicine and Prevention, Tor Vergata University of Rome, *Via* Montpellier, 1, 00133 Rome, Italy; Tel: +390672956164; Fax: +390620427313; E-mail: paola.spitalieri@uniroma2.it

**Keywords:** CRISPR/Cas9, Gene editing, HESCs, hiPSCs, Human diseases, TALENs, ZFN.

## INTRODUCTION

Human embryonic stem cells (hESCs) are pluripotent stem cells obtained from the inner cell mass of embryos at the blastocyst stage [1]. These cells have the unique feature to proliferate unlimited *in vitro*, preserving pluripotent state, moreover they can also differentiate into three germ layers and their derivatives adult tissues. For this reason, hESCs have significant potential for regenerative medicine, even if ethical issues regarding their origin have hindered the clinical application [2]. Thus, fetal, perinatal and adult stem cells can be an alternative source and a new option for regenerative medicine [3, 4]. Unfortunately, immune rejection and differentiation capacities are often not both satisfied by these types of stem cells, making their use inappropriate [5, 6]. An alternative and promising new possibility is the generation of induced pluripotent stem cells (hiPSCs) obtained from adult cells of patients affected by human diseases [7]. hiPSCs play the part of adult and embryonic stem cells, recapitulating the pathological phenotypes and the etiopathology of the diseases *in vitro*. In this way these cells have revolutionized the approaches for understanding the molecular and functional mechanisms of many diseases [8]. Moreover, hiPSCs are a valuable instrument for cell replacement therapy offering an important clinical implication, thanks to their ability to give rise to many disease-relevant cell types. Currently, degenerative diseases are treated with small molecule therapeutics and surgical interventions, looking for alternative therapeutic as iPS technology [9]. In fact the generation of pluripotent cells from patients with developmental or degenerative disorders allows the repopulation of injured or degenerated tissues, thanks to their potential to form *in vitro* relevant somatic lineages [10]. Moreover, patient-specific iPSC-derived systems represent platforms to assess personalized pharmacological therapy of the patient's disease symptoms and to test *in vivo* cell-based repair/modulation of their disease profile. Lastly, disease-specific hiPSCs represent a good target for gene therapy approaches. A new method named 'genome editing' has been recently developed and largely used in the studies of functional genomics, transgenic organisms and gene therapy [11]. Genome editing consists in specific nucleases, which can create *in vitro* site changes in the genomes through a specific DNA-binding domain and a non-specific DNA cleavage domain. Subsequent correction of mutated genes induces established insertions, deletions or substitutions at the loci of interest.

Relatively new approaches have been developed, such as Zinc-Finger Nucleases (ZFNs), Transcription Activator- Like Effector Nucleases (TALENs) and Clustered Regularly Interspaced Short Palindromic Repeats (CRISPR)/CRISPR-associated (Cas) protein 9 system.

ZFNs are engineered nucleases [12], containing a custom Cys2-His2 DNA-binding motif and a DNA catalytic module of the FokI restriction endonuclease. Other popular genome editing platforms are represented by TALENs [13], which are derived from a natural protein of plant pathogenic bacteria Xanthomonas. The DNA-binding domain of TALENs contained between 13 and 28 repeats, each of which is composed of 33–35 amino acids. Two hypervariable amino acids known as the repeat variable diresidues (RVDs) are responsible for DNA binding. Recently, also CRISPR/Cas9 system provides a promising addition to ZFNs and TALENs for genome editing [14]. CRISPR/Cas9 is based on small RNA for sequence-specific cleavage. Because only programmable RNA is necessary to create sequence specificity, CRISPR/Cas9 is easily adaptable and develops very quickly [15, 16]. All the above discussed, gene edit approaches can also be used to induce specific mutations in hiPSCs, converting wild type sequence to mutated one.

In this chapter, we will focus on hiPSCs as a target for gene manipulation. Novel strategies focused to increase the efficiency of genome editing protocols on human reprogrammed cells will be also discussed.

## 1. GENERATION OF HUMAN INDUCED PLURIPOTENT STEM CELLS (HIPSCS)

In 2007, a series of follow-up experiments has been performed in which human adult cells has been reprogrammed into hiPSCs by Shinya Yamanaka's lab. Nearly simultaneously, a research group led by James Thomson achieved the same result. The reprogramming to pluripotent cells has been done through the exogenous induction of four transcription factors such as Octamer- Binding

Transcription Factor 4 (OCT4, also known POUF51) and SOX2 in combination with Krùppel-Like Factor 4 (KLF4) and c -MYC or, alternatively, Nanog and LIN28 [7 - 17]. OCT4 plays an essential role in accomplishing pluripotency for a successful reprogramming. The increased expression of OCT4 respect to the other factors (SOX2, KLF4, and c-MYC) consists in more efficient hiPSCs colony formation, while decreased level of OCT4 reduces the frequency of colonies [18]. Furthermore, induction of the Mesenchymal-to-Epithelial Transition (MET), after addition of bone morphogenetic proteins (BMPs) together to OCT4 alone, can actually bypass the necessity of SOX2, c-MYC, and KLF4 for a successful reprogramming [19].

c-MYC is a strong oncogene, several attempt have been made to supply or exclude this factor. A low reprogramming efficiency was obtained without the adoption of c-MYC but, on the contrary its presence proves to be important for decreasing the tumorigenic potential of hiPSCs [20, 21]. In two studies, Maekawa *et al.* and Nakagawa *et al.* have showed that c-MYC can be replaced with either Glis1, a GLI transcription factor expressed specifically during the growth of embryonic cells, or L-Myc, a transformation- deficient Myc family member, respectively [22, 23]. These factors are also able to generate human and mouse. In the mouse, their use results in iPSCs lines able to produce a chimeric contribution with much less tumor formation and increased germline transmission [22, 23], providing a good set for generating high quality iPSCs. Similarly, KLF4 is only essential during the first step of hiPSCs generation and is considered the trigger of the MET program [19]. Interestingly, however, KLF4 acts also as a transcriptional regulator in the maintenance of pluripotency. Induced expression of KLF4 with either OCT4 or KLF2 lead human ESCs and hiPSCs towards undifferentiated state, thereby significantly increasing the efficiency of colony formation [24], which may facilitate gene targeting in these cells [25]. Lastly, SOX2, associated with OCT4, is believed to inhibit the transforming growth factor beta (Tgf-β) signaling in mouse fibroblast reprogramming [19, 26], and thus can be substituted by Tgf- β signaling inhibitors [27]. Moreover, the quality of hiPSCs depends on the level of expression of the reprogramming factors and on the epigenetic modifications of the somatic cell [28].

Initially, reprogramming transcription factors have been introduced singly using retroviral or lentiviral constructs one for each gene, driving to a high percentage of casual genomic insertions and higher risk of clones partially reprogrammed [17, 29, 30]. Successively, have been used lentiviral polycistronic constructs, delivering set of "reprogramming" factors and reducing the number of genomic integrations. These methods include the use of LoxP sites and Cre-induced excision to eliminate the factors once integrated [30], allowing in this way for the generation of transgene-free hiPSCs [31, 32].

hiPSCs, reprogrammed with this lentiviral method, show greater similarity to human ESCs and possess an improved ability to differentiate without reprogramming transgenes. Other strategies have been subsequently developed to generate transgene-free hiPSCs using protein transduction [33], non-integrating viral vectors such as the Sendai virus [34], episomal vectors [35], transfection of modified mRNA transcripts [36], and chemicals [37]. However, using this protocol, the efficiency of reprogramming become lower (approximately 0.001). A modified mRNA-based method is currently being tested to generate transgene-free hiPSCs [38, 39]. Other strategies utilizing small molecules have also been demonstrated to increase the yield of hiPSCs derivation [24, 40 - 46]. The typical efficiency of hiPSCs generation ranged from 0.01%–5%, depending on the cell source and reprogramming method.

hiPS cells used in clinical therapies should be generated following good manufacturing practice (GMP) conditions, considering invaluable information related to the mechanistic complexity of the reprogramming process.

One of the factors that can strongly influence on the performance of reprogramming is the kind of starting cells, which in turn relies on gene expression level, the state of methylation of the target cells and also the maintaining of pluripotent state. A recent report demonstrated that cell type of origin minimally contributes to hiPSC variability and more importantly that epigenetic differences existing among hiPS lines are only in minimal part due to the starting cell type [47]. This data confirms the results already suggested in another study published two years ago [48]. Authors reported that hiPS lines obtained from the same donor are highly similar to each other. The genetic variations are responsible of a donor-specific expression and methylation profile in reprogrammed cells that in turn modulate their differentiation potentiality. Moreover, the culture conditions might also concur to the differences obtained between the several hiPS cell populations generated. Finally, some changes may be associated to stochastic events during reprogramming, which cannot be predicted [49]. However, human iPS cells and embryonic stem cells although having a different origin are biologically very similar, having the same capability to give rise to all somatic lineage of the body.

The capacity to dedifferentiate patient-specific cells into pluripotent stem cells and to re- differentiate them into cells representative of the disease, organ or tissue, allows to faithfully reproduce the key aspects of the disease in a 'petri dish', to quantify disease progression and/or regression in different tissues and to open up new perspectives in the field of drug discovery [50]. hiPSCs, *in vitro* differentiated into relevant adult cells, are used both for the assessment of appropriate pharmacological management of the patient's disease symptoms and for the *in vivo* cell-based repair or modulation of their disease profile. hiPSCs represent a robust source of progenitors for regenerative personalized medicine, overcoming immunological rejection insofar as patient-specific. It has been agreed that hiPSCs represent a convenient tool to examine the pathogenesis and the progression of human disease avoiding trials based on conventional rodent and cell lines [51]. In fact, murine models of human congenital and acquired diseases are helpful systems but human pathophysiology cannot be always faithfully reproduced. When animal and human physiology are different, disease-specific stem cells able to differentiate into the disease target cells, that can be genetically corrected in a specific way for each genetic defects.

The benefit of using hiPSCs -based models are listed below:

- hiPSCs can be derived from adult somatic cells, embryonic/fetal cells, adult stem cells, cancer cells;
- when derived, hiPSCs naturally conserve parental genetic background;
- hiPSCs have the capability to differentiate into a several cell types in culture;
- hiPSCs can self-renew and maintain their stemness;
- hiPSCs mimic *in vitro* early human embryonic development ongoing differentiation.

For all these reasons, hiPSCs represent a helpful model system particularly in those cases where animal models do not adequately reproduce human phenotype or when disease-target cells types are not accessible for research.

Moreover, hiPSCs for their self-renewing capability can be gene-targeted, cloned, genotyped, and expanded. The possibility of hiPSCs to be corrected or modified genetically offers extraordinary clinical promises for generating artificial organs and safer gene therapies.

## 2. GENE EDITING

Technologies for manipulating DNA have enabled advances in biology ever since the discovery of the DNA double helix, when scientists have been contemplating the opportunity of bringing site-specific changes to cell and organism genomes.

Genome editing is based on the use of engineered nucleases composed of sequence-specific DNA-binding domains fused to a non-specific DNA cleavage module [52, 53]. These nucleases induce targeted DNA double-strand breaks (DSBs) that stimulate the cellular DNA repair mechanisms, such as error-prone non-homologous end joining (NHEJ) and homology-directed repair (HDR) [54]. More recently, the development of sequence-specific nucleases, such as Zinc Finger Nucleases (ZFN), Transcription Activator-like Effector Nucleases (TALEN), or CRISPR/Cas9 nucleases have enhanced our possibility to engineer genetic modifications in human cells [52, 55 - 57].

These nucleases can be custom-engineered in order to stimulate DSBs at a specific target site within the genome. If these breaks are repaired using the non-homologous end joining (NHEJ) pathway, small insertion and deletion mutations (indels) are produced, disrupting genes. Alternatively, the DSBs can be repaired by the homologous recombination pathway, in this case specific base pair changes or gene insertions can be formed using a homologous donor targeting vector (Fig. **1A**). These nucleases use identical mechanism of action: they generate chromosomal DNA break in a site-specific manner, stimulating endogenous DNA repair systems that consequently result in targeted genome modifications. Anyway, each nuclease has exclusive peculiarity. The possibility to generate patient-specific iPSCs together with the advent of engineered nucleases, developed for gene modification, surely consist in a good opportunity for cell therapy of several inherited genetic diseases.

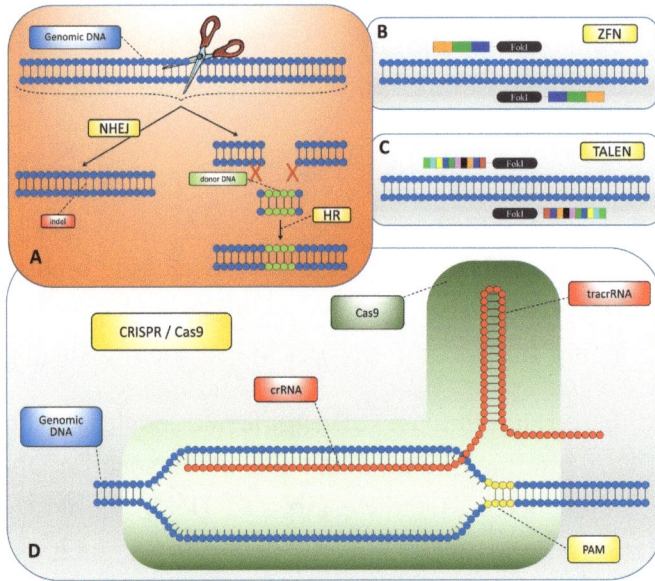

**Fig. (1). (A)** Nuclease mediated Double Stranded Breaks (DSB) DNA induce Targeted Gene Inactivation or Homologous Gene Replacement. If targeted nucleases are co-transfected with a homologous donor DNA fragment, homologous recombination (HR) substitutes altered DNA with a corrected sequence (right). In the lack of a DNA donor fragment, non-homologous end joining restores (NHEJ) the break, but with frequent insertions and deletions (indel), thus inactivating the gene.
**(B)** Zinc-Finger Nucleases (ZNF) work as dimers to cut double strand DNA: pairs of ZFNs are designed to bind to adjacent sites in the genome to allow *FokI* dimer formation and double stranded DNA cleavage.
**(C)** TALENs bind as dimers to cut double strand DNA: TALENs include multiple repeat variable diresidues (RVDs) which specifically bind to a single nucleotide and fused to non- specific cleavage domains such as *FokI*.
**(D)** CRISPR/Cas9 is an RNA guided Genome Engineering System: consist of a CRISPR- associated (Cas9) endonuclease that assembles with two small guide RNAs, crRNA and tracrRNA, to create a double-stranded DNA break (DSB) in a sequence specific manner. The crRNA and tracrRNA, forming a single guide RNA (sgRNA), directs the Cas9 nuclease to the target sequence through base pairing between the sgRNA sequence and the genomic target sequence. The target sequence consists of a 20-bp complementary to the sgRNA, followed by trinucleotide sequence (5'-NGG-3') called the protospacer adjacent motif (PAM). The Cas9 nuclease digests both strands of the genomic DNA 3-4 nucleotides upstream the 5'- terminus of the PAM sequence. Through a simple change of the guide RNA sequences, the Cas9 can be designed to introduce site-specific DNA double-strand breaks virtually anywhere in the genome.

Thus, an efficient genetic manipulation technology on hiPSCs can revolutionize the way to investigate the cellular and developmental functions of human genes.

Homologous recombination allows targeted gene inactivation, substitution or also insertion; however, the limited usefulness of this method has been limited by the low efficiency in mammalian cells and animal models.

Considering that induction of a DSB increases the rate of HDR, targeted nucleases have become the tool of choice for stimulating the efficiency of HDR-mediated genetic modifications. By introducing a site-specific nuclease with a donor DNA template homologous to the target locus [58], single or multiple transgenes can be efficiently delivered in a specific manner [59].

Linear donor sequences with <50 base pairs of homology [60], as well as single-stranded DNA oligonucleotides [61], can specifically generate mutations, deletions or insertions at the target site.

Several gene editing strategies have proved to be applicable to correct several defective genes in hiPSCs. The selection of the protocol depends on the gene correction method and the mutation type. Gene corrected-iPSCs can be then differentiated into suitable somatic cells before delivering them to patients. *In vitro* analyses can also be performed in order to evaluate the expression of the corrected gene and in the same time to avoid the risk of teratoma formation *in vivo*. ZFNs and TALENs use the principles of DNA protein recognition and thus are site-specific developed, while recently several groups report the latest class of genome editing nucleases Cas (CRISPR associated), the specificity of which is mostly determined by small guide RNAs rather than by DNA-binding proteins [62].

## 2.1. Zinc Finger Nuclease (ZFNs)

A Zinc Finger Nuclease is composed of two domains: a DNA-binding Zinc Finger Protein (ZFP) domain and the nuclease domain derived from the FokI restriction enzyme (Fig. **1B**).

Some research groups have indicated nucleases for genome engineering of hiPSCs by linking the cleavage domain of the FokI restriction enzyme to a designed Zinc Finger Protein (ZFP). Thus, two ZFN monomers make an active dimeric structure; each monomer is separated from the other by a spacer of 5-7 bp, allowing the formation of dimers and providing more specific targeting ability of ZNFs. The FokI dimeric interface is engineered to work as an obligate heterodimeric forms, such that reduces off-target effects and ZFN cytotoxicity [63, 64]. ZFPs consist of tandem arrays of C2H2 zinc-fingers, the most common DNA-binding motif in higher eukaryotes, that strongly increases the specificity of the sequence [65]. Each Zinc Finger binds to 3-bp DNA sequence and 3–6 ZF arrays are necessary to create a single ZFN subunit that pairs up to DNA sequences of 9–18 bp [66]. Crucially, the DNA binding specificities of zinc-fingers can be changed by mutagenesis, which is a key feature of constructing a programmable nuclease [67]. Although each Zinc Finger identifies three DNA bases, a collection of 64 zinc-fingers that recognizes all possible combinations of triplet sites don't exist [68]. Furthermore, not all newly assembled ZFNs, especially those with three zinc-fingers, can cleave chromosomal DNA efficiently; successful target sites are often rich in guanines and consist of 5′-GNN-3′ (where N represents any nucleotide) repeat sequences. Thus, a single functional ZFN pair can be accomplished per ~100-bp DNA sequence on average [69]. The ZFN method is fast and convenient but has a poor targeting density and often lacks of targetable sites for genome editing in small DNA sequences.

In Table **1** are reported some applications of ZFN protocol to disease specific hiPSCs.

**Table 1. ZNF Gene editing in hiPSCs –disease models.**

| ZFN | REPROGRAMMING METHOD | CELL TYPE | CORRECTED DISEASE | REFERENCES |
|---|---|---|---|---|
| | *Lentiviral vectors* (OSKM) | HF | Sickle Cell Disease | *Sebastiano V et al.*, 2011 |
| | *PiggyBac* transposon vector (OSKM + Lin28) | MSCs | | *Zou J et al.*, 2011 |
| | *Episomal plasmids* (OSKM, NANOG, LIN28, SV40LT, hTERT) | HF | α-Thalassemia | *Chang C.J et al.*, 2012 |

*(Table 1) contd.....*

| ZFN | REPROGRAMMING METHOD | CELL TYPE | CORRECTED DISEASE | REFERENCES |
|---|---|---|---|---|
| | *Lentiviral vectors (OSKM)* | HF | Parkinson's Disease | *Soldner F et al.*, 2011 |
| | Provided by George Q. Daley | HF | Down Syndrome | *Jiang J et al.*, 2013 |
| | *Moloney murine leukemia virus-derived vectors (OSKM)* | HF | $\alpha_1$-Antitrypsin Deficiency | *Yusa K et al.*, 2011 |
| | *Retroviral vectors (OSKM, NANOG)* | HF | Cystic Fibrosis | *Crane et al.*, 2015 |
| | *Retroviral vectors (OSKM)* | MSCs | X-Linked Chronic Granulomatous Disease | *Zou J et al.*, 2011 |
| | *Lentiviral vector (OSKM)* | CD34+cells | | *\*Dreyer A.K et al.*, 2015 |

O: OCT4; S:SOX2; K:KLF4; M:c-MYC; HF: Human Fibroblast; MSCs: Mesenchimal Stem Cells.

\* Dreyer A.K *et al.*, compare Zinc Finger Nucleases (ZFNs) efficiency to TALENs.

## 2.2. TALENS

An alternative nuclease for genome editing can be designed using a simple modular DNA recognition code by Transcription Activator Like Effector (TALE) proteins, secreted by Xanthomonas plant pathogens [70, 71]. The DNA binding domain is composed of multiple 34 amino acid repeats, organized in tandem (Fig. **1C**). TALE specificity is guaranteed by two hypervariable amino acids that are called "repeat-variable diresidues" (RVDs) [72, 73]. TALE repeat domain is responsible to bind to individual bases in a target-binding site, so any desired genomic sequence can be recognized as a DNA-binding domain. Nucleases based on such engineered TALE domains can introduce targeted alterations in endogenous genes within transformed human cells.

Like Zinc Fingers, modular TALE repeats are assembled each other to identify contiguous DNA sequences. However, in contrast to Zinc Finger Proteins, there are no re-engineering of the linkage between repeats necessary to construct long arrays of TALEs, allowing the possibility to address single sites in the genome. While the single base recognition of TALE-DNA binding repeats provides greater design flexibility than triplet limited Zinc Finger Proteins, the cloning of repeat TALE arrays presents an important technical challenge due to extensive identical repeat sequences. To overcome this problem, numerous methods have been evolved that enable rapid design of custom TALE arrays.

Several studies using various assembly strategies have explained that TALE repeats can be linked to identify any user-selected sequence [74, 75]. The only targeting impediment for TALE arrays is that TALE binding site must begin with a T base.

However, a recent study reports the construction of a library of TALENs targeting 18,740 human protein-coding genes [76], promoting new challenges [59]. Several studies have suggested the benefit of TALENs [77, 78] in hiPSCs for gene editing (Table **2**).

Table 2. TALEN Gene editing in hiPSCs –disease models.

| TALEN | REPROGRAMMING METHOD | CELL TYPE | CORRECTED DISEASE | REFERENCES |
|---|---|---|---|---|
| | *Lentiviral vector (OSKM)* | CD34+cells | X-Linked Chronic Granulomatous Disease | *Dreyer A.K et al.,* 2015 |
| | Provided by Dr. M. Wernig | n.r. | Sickle Cell Disease | *Sun N et al.,* 2013 |
| | *Retroviral vectors (OSKM)* | HF | Recessive Dystrophic Epidermolysis Bullosa | *Osborn M.J et al.,* 2013 |
| | *Retroviral vectors (OSKM)* | n.r. | $\alpha_1$-Antitrypsin Deficiency | *Choi S.M et al.,* 2013 |
| | *Episomal vector (O, S, K, SV40LT, miR-302–367)* | AF | β-Thalassemia | *Ma N et al.,* 2013 |
| | *Lentiviral vectors (OSKM)* | HF | Niemann-Pick Type C | *Maetzel D et al.,* 2014 |
| | *Lentiviral vector (OSKM, NANOG, LIN28)* | BM-MSC | X-linked Severe Combined Immunodeficiency | *Menon T et al.,* 2015 |
| | *Retroviral vectors (OSKM)* | HF, K | Cystic Fibrosis | *Camarasa M.V et al.,* 2016 |

*(Table 2) contd.....*

| TALEN | REPROGRAMMING METHOD | CELL TYPE | CORRECTED DISEASE | REFERENCES |
|---|---|---|---|---|
| | *Episomal vectors (OSKM)* | HF | Hemophilia A | *Park C.Y et al.*, 2014 |
| | *Episomal vectors (pCXLE-hOCT3/4-shp53-F, pCXLE-hSK, pCXLE-hUL)* | HF | Duchenne Muscular Dystrophy | *Li H.L et al.*, 2014 |

O: OCT4; S:SOX2; K:KLF4; M:c-MYC; HF: Human Fibroblast; AF: Amniotic Fluid; n.r.: not reported; BM-MSC: Bone marrow multipotent stem cells

\* Dreyer A.K *et al.*, compare Zinc Finger Nucleases (ZFNs) efficiency to TALENs.

\* Li H.L *et al.*, use both TALENs and CRISPR/Cas9

## 2.3. CRISPR/Cas9

In the mid-2000s a few microbiology and bioinformatics laboratories started investigating CRISPRs (Clustered Regularly Interspaced Palindromic Repeats), already described in the genome of *Escherichia coli* [79] and in numerous Bacteria and Archaea [80]. In 2005, it was reported that variable sequences, called "spacers", within CRISPRs were shared with plasmid and viral vectors [81 - 83] and also that CRISPR loci and Cas (CRISPR-associated) genes codify proteins with presumptive nucleolytic and helicase activity [84 - 86]. In 2007, CRISPR/Cas mediated adaptive immunity has been demonstrated by invasion experiments of the lactic acid bacterium Streptococcus Thermophiles [87]. Probably CRISPR/Cas represents an immune defense mechanism using antisense RNAs as memory of past infections [88]. Cas9, a CRISPR-associated endonuclease, can be directed to specific DNA sequences to cause double-strand breaks thanks to the presence of the trans-activating CRISPR RNA (tracrRNA): CRISPR RNA (crRNA) duplex. A single guide RNA (sgRNA) for genome engineering can be formed by the union of crRNA-tracrRNA hybrids pair to the 3'-end double-stranded structure binding Cas9. The sgRNA could facilitate CRISPR-Cas 9 to any target DNA sequence with a protospacer-adjacent motif (PAM), by changing the guide RNA sequences (Fig. **1D**) [56, 89].

In 2012, the S. Pyogenes CRISPR-Cas9 protein has been described as a dual-RNA–guided DNA endonuclease that employs the tracrRNA:crRNA duplex [90] to direct genome engineering [91]. Three CRISPR/Cas system types (I, II, and III) utilize specific molecular mechanisms to obtain nucleic acid identification and cut [92]. The type I and type III systems utilize a large complex of Cas proteins for crRNA-guided targeting [93 - 95]. However, the type II system appears extremely helpful for genome engineering applications, requiring only a single protein for RNA-guided DNA recognition and cleavage [91, 96]. Several delivery approaches can be employed to *in vitro* vehicle plasmid DNA translating Cas9-gRNA complexes through cell membranes in cell cultures: electroporation, nucleofection and lipofectamine-mediated transfection [15, 56, 97]. Cas9 (formerly COG3513, Csx12, Cas5, or Csn1) has been identified as a large multifunctional protein [80] with two hypothetical nuclease domains, HNH and RuvC-like [86 - 88]. Cas9 is necessary for the response against viral invasion, introducing DSBs into foreign plasmids and phages following their invasion [98] and interfering with plasmid transformation efficiency [99]. In particular, Cas9 binds the open chromatin rather than areas of compact, transcriptionally inactive chromatin [100].

In contrast to ZFNs and TALENs, which require substantial protein engineering for each DNA target site to be modified, by introducing the CRISPRCas9 system has been simplified the recognition mode by developing the variable guide RNA sequence.

In 2013, two studies simultaneously showed the derivation of CRISPR type II systems from Streptococcus thermophiles [89] and Streptococcus pyogenes [56, 89] to realize genome editing in mammalian cells. Heterologous expression of mature crRNA-tracrRNA hybrids, as well as sgRNAs [56, 89], conducted Cas9 cleavage within cellular genome to induce NHEJ or HDR- mediated genome editing. Multiple guide RNAs are able to target different genes at once. Some reports showed that CRISPR-Cas9–mediated editing efficiency can reach 80% or more, depending on the target [101, 102].

Other examples of CRISPR-Cas9 applications with relevance to human health consisted in the capacity to edit genetic mutations that cause inherited disorders. These studies underlined the possible use of this technology for human gene therapy to cure genetic disorders [103].

In Table **3** some examples of CRISPR/Cas9 technology applied to models resembling genetic disease are listed.

**Table 3. CRISPR/Cas9 Gene editing in hiPSCs –disease models.**

| CRISPR/Cas9 | REPROGRAMMING METHOD | CELL TYPE | CORRECTED DISEASE | REFERENCES |
|---|---|---|---|---|
| | *Episomal vectors* (pCXLE-hOCT3/4-shp53-F, pCXLE-hSK, pCXLE-hUL) | HF | Duchenne Muscular Dystrophy | *\*Li H.L et al.*, 2014 |
| | *Lentiviral vectors (OSKM,Nanog,LIN28)* or *Sendai virus (OSKM)* | HF | Cystic Fibrosis | *Firth A.L et al.*, 2015 |
| | *Lentiviral vectors (OSKM)* | HF | X-Linked Chronic Granulomatous Disease | *Flynn R et al.*, 2015 |
| | *Episomal vectors* Or *Sendai virus* | Urine cells | Hemophilia A | *Park C.Y et al.*, 2015 |
| | *Lentiviral vectors (OSKM)* | K | Severe Combined Immunodeficiencies | *Chang C.W et al.*, 2015 |
| | *Episomal vectors* | hHC | Sickle Cell Disease | *Huang X et al.*, 2015 |
| | *Sendai virus (OSKM)* | HF | β-Thalassemia | *Xie F et al.*, 2014 |
| | *n.r* | n.r | $\alpha_1$-Antitrypsin Deficiency | *Smith C et al.*, 2015 |

O: OCT4; S:SOX2; K:KLF4; M:c-MYC; HF: Human Fibroblast; K: Keratinocytes; hHC: human Hematopoietic cell.
\* Li H.L *et al.*, use both TALENs and CRISPR/Cas9.

In this chapter, we describe recent applications of gene therapy approaches by using the above mentioned endonucleases in human induced pluripotent stem cells carrying mutations responsible of monogenic diseases.

Two studies have been published reporting the genetic correction of hiPSCs from patients with Sickle Cell Disease (SCD; OMIM #603903), the most common human genetic disease affecting more than 300,000 individuals worldwide each year [104 - 106]. The disease is caused by a homozygous point mutation (from A to T) in the sixth codon of the human β-globin (HBB) gene, resulting in an altered form of adult hemoglobin in oxygen-carrying red blood cells. The first report shows specific ZFN-mediated gene editing of the point mutation in the HBB locus. A donor vector with a loxP-flanked ('floxed') hygromycin resistance selection cassette is used, allowing the detection of hygromycin-resistant clones.

After excision of selection gene cassette, the corrected hiPSCs are differentiated into erythroid cells, expressing restored HBB transcript [105]. The other gene correction approach for the sickle cell mutation follows a similar selection based approach [104]. Authors achieve an optimal correction of the mutation in human iPS cells derived from patient affected by sickle cell anemia employing ZFNs engineered by the publicly available Oligomerized Pool ENgineering (OPEN) method, already described [107]. By inserting a drug-resistance cassette into a neighboring intron up to 82 base pairs away from the ZFN cut site the simultaneous correction and selection of the modified cells are obtained.

Following removal of the factors responsible for pluripotency and of the drug-resistance cassettes, restored hiPSCs *in vitro* differentiated into cells belonging to the three germ layers.

Sickle cell disease hiPSCs are also employed as model to edit the mutations provoking disorder by TALENs with high specificity and modularity. Using this approach, authors construct a pair of custom-developed TALENs to identify the mutation site at the human β-globin (HBB) gene together with the donor vector, containing a piggyBac DNA transposon. After TALEN-mediated gene correction, hiPSCs maintain their pluripotent state both *in vitro* and *in vivo*. In addition, TALEN –mediated gene correction doesn't produce any evident off-target event or chromosomal modification, showing in this way the advantage of this method for the treatment of genetic diseases such as SCD [108].

Recent studies use CRISPR/Cas9 to develop better designer nucleases for gene targeting in human iPSCs. Huang X. and colleagues have utilized a specific guide RNA together with Cas9 to correct one allele of the SCD HBB gene in human iPSCs by homologous recombination using a donor DNA template that carry the wild-type HBB DNA and a selection excisable cassette. They demonstrate that the corrected gene in the erythrocytes obtained from genome-edited iPSCs is able to express wild type HBB protein after differentiation, highlighting the efficiency and the specificity of CRISPR/Cas9-mediated genome editing [109].

Also β-thalassemia (β-Thal; OMIM #613985) always caused by HBB mutations, can be treated by gene editing approach. β-Thal is an autosomal recessive blood disorder due to a reduced or absent synthesis of β-globin chains [110]. Clinical signs of the affected patients consist of severe anemia and a shortened life span. For correcting β-Thal mutations, a donor template has been designed, inside TALENs targeting site, containing the whole wild type HBB gene. This template could serve to substitute the endogenous gene with aberrations in any site within the HBB region. Using a reporter assay, authors demonstrate that the cleavage activity of the constructed TALENs is highly specific. The gene-corrected β-Tha--hiPS lines can be stimulate to differentiate both into hematopoietic progenitor cells and into erythroblasts. This work demonstrates that the editing of disease-causing mutations in β-Thal hiPSCs could rescue physiological expression of full-length β-globin [111]. Recently has been reported the site-specific correction in the HBB locus in β-thalassemia hiPSCs associating the CRISPR/Cas9 approach together with the piggyBac system. The combination of CRISPR/Cas9 and the piggyBac represents an ideal system to cleave the HBB gene and in the same time to select for homologous recombination events. HBB gene correction has been confirmed by a restored expression of the protein in erythroblasts differentiated from hiPSCs [112]. Both β-thalassemia and sickle cell disease are included in "haemoglobinopathies", that display defects of mature haemoglobin. In turn haemoglobinopathies belong to the group of bone marrow diseases that also comprise primary immunodeficiencies (PIDs). PIDs are an heterogeneous set of inherited disorders responsible for dysfunctions of the immune system, including X-linked Severe Combined Immunodeficiency (SCID-X1; OMIM #300400), Adenosine deaminase deficiency-SCID (ADA-SCID; OMIM #103700), and Wiskott- Aldrich syndrome (WAS; OMIM #301000).

Severe combined immunodeficiency is the most severe form of PID, patients show severe defects in T cells with or without accompanying defects in B cells. The most common T-B+ immunodeficiency mutations occur in the Janus kinase-3 (JAK3; #600173) gene (autosomal recessive) and in the X-linked common gamma chain (γC) gene (IL2RG; #308380) [113]. SCID (OMIN #600802) due to JAK3 mutation segregate as autosomal recessive and is characterized by the absence of circulating T and natural killer (NK) cells and by a normal number of poorly functioning B cells (T– B+ NK-).

Chang *et al.*, demonstrate that differentiation of JAK3-deficient human T cells is blocked at an early developmental stage, while editing JAK3 mutation by using CRISPR/Cas9-enhanced gene targeting rescue the differentiation potential of early T cell progenitors. These corrected progenitors are able to generate NK cells and mature T cell populations [114]. Mutations in the gene translating the interleukin 2 receptor common gamma chain (IL2RG) are responsible of diseases such as severe combined immunodeficiency, x-linked, T cell-negative, B cell-positive, NK cell- negative (SCID-X1). Menon *et al.*, describe the correction of SCID-X1 hiPSCs utilizing TALENs. The mutation leads to an aberrant splicing of the IL2RG transcript, determining the defect of normal NK and T cells. SCID-X1 mutation has been corrected by TALEN pairs together with a donor plasmid, carrying the corrective DNA sequence. The differentiation of modified hiPSCs into mature NK cells and T cell precursors has been reported by mature cell markers expressing a correctly spliced IL2RG transcript [115]. Also X-Linked Chronic Granulomatous Disease (X-CGD; OMIM #300481) mutations have been corrected by ZNF. The disease is determined by alterations in the CYBB gene encoding the transmembrane gp91phox subunit of phagocyte NADPH oxidase, required for micro biocidal reactive oxygen species (ROS) production by neutrophils and monocytes [116, 117]. The correction has been performed inducing a sequence-specific double-strand DNA break and inserting a normal copy of the CYBB gene in the adeno-associated virus integration site 1 (AAVS1) safe harbor locus. Corrected clones have been selected and full correction has been reported in neutrophils differentiated from X-CGD hiPSCs, that fully modelled the oxidase- negative disease phenotype [118].

The same mutation restoration has been reported by Dreyer *et al.*, comparing AAVS1-specific Zinc Finger Nucleases (ZFNs) efficiency to TALENs. The approach contemplates the integration of microRNA-223 promoter driven gp91phox transgene to the AAVS1 site of X-CGD patient-derived hiPSCs, restoring in this way the function of the NAPDH oxidase (NOX) complex. Previously, ZFNs have been already exploited in CGD-hiPSC to mediate functional correction using a constitutively active CAG promoter-driven therapeutic gp91phox expression cassette at the AAVS1 locus [118]. However, authors employ the myeloid-specific microRNA-223 promoter to confine gp91phox expression to myeloid cells, avoiding harmful consequences of constitutive expression of gp91phox in the stem cell compartment. Myeloid differentiation of edited hiPSCs show normal gp91phox expression levels compared to wild type cells. Moreover, both restoration of ROS production and generation of neutrophil extracellular traps (NETs, composed of chromatin and granule proteins that are able to recognize and destroy extracellular microorganism, corroborating the validity of gene targeting) has been achieved, demonstrating the functional reading of the cellular pathological phenotype. The results showed that TALEN pair leads to the generation of correctly targeted hiPSC clones more efficiently than ZNFs [119].

Lately CRISPR-Cas9 has been employed by Flynn *et al.,* always to correct a single intronic CYBB mutation. Gene modification has been achieved by the restoration of a skipped exon in the cytochrome b-245 heavy chain (CYBB) protein, determining the repair of oxidative burst function in phagocytes obtained from hiPSCs in a highly efficient manner [120].

Programmable nucleases have been also applied to correct the most common mutation responsible of Cystic Fibrosis disease (CF; OMIM # 219700). CF is an autosomal recessive disorder; the main defect is represented by a dysregulation of epithelial chloride transport caused by mutations within the CF transmembrane conductance regulator (CFTR) gene [121]. Loss of function in the CFTR in organs with secretory function is a significant cause of this multisystem disease. Lung infections are very present and responsible for 80%-90% of the deceases in CF patients [121]. No resolutive treatments are available, thus the transplantation of CFTR- corrected, autologous lung stem/progenitor cells provides an attractive alternative strategy to cure the disease. Patient-specific hiPS cells, carrying ΔF508 mutation, has been corrected designing a homologous recombination donor vector which includes a PiggyBac transposon- based double selectable marker cassette in combination with TALENs [122]. The restoration of functional CFTR channel has been demonstrated by airway and intestinal epithelial cells derived from edited pluripotent stem cells. The protocol allows the correction of patient-specific hiPSCs in about 3 months: moreover, repaired cells can be cultured and expanded, representing a useful resource for therapeutic applications.

Respect to ZNF approach, TALEN displays several advantages, considering that repair vector has been designed as a transposon with a double selectable marker, which promotes the system of selection and seamless exclusion of the selection cassette.

Also Zinc Finger Nuclease (ZFN)-mediated homology-directed repair (HDR) has been used for correcting CFTR mutations either one or both mutant alleles in hiPSCs derived from either compound heterozygous DI507/DF508 or homozygous DF508/DF508 genotypes ZFNs targeting, designed for exon 10, recognizes DNA sequences approximately 110 bp upstream of either the ΔI507 or ΔF508 deletion and has been co-delivered with a plasmid encoding CFTR. Corrected CF-hiPSCs reveal the functional restoration of the mature CFTR glycoprotein and of its function after differentiation into epithelial cells [123]. At least CF hiPSCs, homozygotes for ΔF508 mutation, have been corrected also using a customized CRISPR system, containing two components: a plasmid encoding the full-length Cas9 protein and another plasmid carrying a gRNA cassette constructed to target sequences near to ΔF508 mutation in the endogenous CFTR gene. The results have showed that the mutation has been exactly corrected, as demonstrated by the recovery of normal CFTR expression and function in mature differentiated airway epithelial cells [124]. Although the three different systems are extensively used to induce efficient homologous recombination, some important aspects have to be considered. ZFNs are difficult to design and also economically expensive to obtain, while TALENs can be produced quickly (almost one week) using an established protocol, but numerous pairs have to be created to comparatively evaluate their cutting efficiencies. In addition, TALENs ability can be hampered by DNA methylation and histone acetylation in inactive chromatin [12, 125 - 127]. On the other side, the Cas9 nuclease does not have these limitations, is readily disposable, and is now widely used to correct disease-causing mutations in hiPSCs. An elegant approach has been reported by Yusa *et al.*, using hiPSCs derived from patient affected by α1-Antitrypsin Deficiency (A1ATD; OMIM # 613490), an autosomal recessive disorder, caused by a single point mutation in the A1AT gene (the Z allele; Glu342Lys) [128, 129]. Authors showed that footprint-free genomic correction can be obtained applying ZFNs and piggyBac DNA transposon as an excisable selection cassette. The piggyBac selection cassette has been subsequently eliminated from the site of correction with the transient expression of Piggy-Bac transposase, showing that no trace of exogenous sequences remained at the targeted locus.

Differentiation *in vitro* of hiPSCs into hepatocyte-like cells, the target cell type involved by A1ATD, demonstrates expected phenotypic recovery following the genetic modification of hiPSCs [130]. Moreover, a comparable or even higher gene targeting efficiency has been obtained by choosing the TALEN technology which adopts a pair of designed and constructed TALEN expression vectors, able to detect the flanking sequences of A1AT mutation in combination with previously reported donor construct [130]. TALEN corrected hiPSCs differentiate into mature hepatocyte-like cells, exhibiting a normal metabolic function, as demonstrated by the activities of four major cytochrome P450 enzymes. The results prove TALEN as an efficient, robust and economic in alternative to ZFN technology [131]. A parallel study has experimented the CRISPR/Cas9 approach on the same cells: significantly higher frequencies of non-homologous end joining (NHEJ)-mediated insertions/deletions has been reached at several endogenous loci respect to those observed using TALENs. These comparative results suggest the ability of Cas9 of specifically recognizing and binding a single-nucleotide mutation or variant in human iPSCs [132]. In a recent study, mutations /deletions in the dystrophin gene leading to a lack of exon 44 has been targeted both by TALEN and CRISPR-Cas9 gene editing systems, restoring the expression of the full-length human dystrophin in Duchenne Muscular Dystrophy (DMD) patient-derived iPSCs [133]. DMD (OMIM #310200), X-linked genetic degenerative myopathy, is the most prevalent congenital pediatric muscular dystrophy. The major clinical manifestations of this multisystem disease included disease-specific serological abnormalities, dilated cardiomyopathy, cataracts, insulin-resistance, cardiac conduction defect, myotonia and muscular dystrophy, which can led to motor function loss during puberty [134]. The disease is determined by mutations in the dystrophin gene (locus Xp21.2), causing dystrophin loss of impairment. The disease is caused by a decrease of dystrophin, a protein binds to the dystrophin glycoprotein complex (DGC), linking the cytoskeleton to the extracellular matrix in skeletal and cardiac muscles [135]. The deletion of exon 44 is one of the most common deletion in DMD patients, however the cleavage of each deletion differs among patients. In order to rescue the full amino acid sequence of the dystrophin protein for DMD patients who lack exon 44, three different methods has been designed: skipping exon 45 by disruption of its splicing acceptor, small indel-mediated frameshift of exon 45, and exon 44 knock-in. Both TALEN and CRISPR-Cas9 strategies has been tested. To confirm the genetic correction, hiPSCs has been differentiated into skeletal muscle cells, evaluating the expression of the dystrophin protein.

The results demonstrate that all three methods are able to restore the protein expression in differentiated skeletal muscle cells, even though the exon knock in system results the most effective. Regarding the gene editing protocol, TALEN and CRISPR are equally efficient, having small consequences on off-target mutagenesis when targeted to a single sequence region [133]. Finally, both TALEN and CRISP have been experimented by the same group for targeting hiPSCs derived from patients with Hemophilia A (OMIM #306700), an X-linked genetic disorder, with an incidence of 1 in 5,000 males worldwide. Mutations in the F8 gene, codifying for the blood coagulation factor VIII, are responsible for disease.

Approximately 50% of severe hemophilia A cases are due to two different kind of chromosomal inversions that affect a part of the F8 gene [136 - 138]. Actually, there is no cure for Hemophilia A and genome editing using programmable nucleases offers good opportunity to treat the disease. TALENs have been used to introduce the 140-kbp chromosomal inversion in wild type hiPSCs and soon after to restore the same micro rearrangement using the same TALEN pair. F8 wild type expression has been evaluated in endodermal and endothelial cells derived from genome- corrected hiPSCs [139]. Successively, another paper reports the application of the type II clustered CRISPR)/CRISPR-associated (Cas) system to revert two recurrent chromosomal inversions (140-kbp or 600-kbp) causing severe hemophilia A. Phenotypic rescue has been established both *in vitro* and *in vivo* in differentiated endothelial cells. This report firstly demonstrates that chromosomal inversions or other large rearrangements can be corrected using CRISPR-Cas9 system in patient hiPSCs [140].

## 3. PERSONALIZED MEDICINE

The field of biomedical sciences has been profoundly changed through the hiPSCs reprogramming technology. As patient-specific hiPSCs make accessible tissues available for *in vitro* studies and drug screening experiment, the paradigm of personalized medicine gets a new scope and the field is now rapidly moving towards clinical applications. Novel experimental studies are nowadays able to map the molecular differences between diseased and healthy samples to generate unique insight into mechanisms of clinically defined diseases and hypothesize novel targets for therapeutic intervention. Human iPS-disease-specific lines can be used for drug development and screening. Drug treatment can be tailored to patients from whom hiPSCs are derived. The use of hiPSCs for evaluation of efficacy and toxicity of drug compounds will also improve drug toxicology analysis. hiPSCs provide an unlimited source of primary cells useful to examine the pathways leading to disease pathogenesis at the cellular level. Animal models have a number of inherent flaws in their use for toxicity assays: they may not compare well with human physiology, are ethically contentious and are expensive to acquire and sustain, respect to *in vitro* cultured cells. hiPS lines can be obtained from large populations of individuals with a wide species of well-characterized phenotypes, and can be used in a high throughput experimental analysis. Thus, they can serve as useful and valuable tools for studying how variation toxic susceptibility correlates to genetics, disease progress and other observable phenotypic characteristics [141]. Most human cell lines that are currently in use for biomedical research and drug development are immortal cell strains with genetic modifications and epigenetic abnormalities, and are developed from either malignant tissues or from a genetic manipulation process that derives immortal expansion. Primary human cells are regulatory used, though their life span in culture is limited. Finally for many diseases, a suitable human experimental model does not currently exist and disease-specific hiPSCs can been generated to improve the study of molecular pathways and therapeutic agents for human diseases [142, 143]. All these data suggest that such unique humanized iPSCs-based disease-model systems exploit an individualized cell-based platform that has unlimited growth potential for novel regenerative strategies and clinical therapeutics along with companion diagnostics to predict and prognosticate the molecular basis of several human diseases. It is evident that developments in our knowledge of cellular reprogramming will allow the application of patient specific cells in therapy, for the exploration of novel diagnoses parameters, or for the choice and the discovery of innovative molecules with therapeutic potential.

Any advancements will enable researchers and clinicians to convert corrected patient specific cells in treatment options.

In conclusion, reprogramming technologies are a reality for the modeling of disease, identification and testing of innovative therapies. Future advantages will concur to cell transplantation strategies as potential therapeutic solutions for the improvement of human health.

## CONFLICT OF INTEREST

The authors confirm that they have no conflict of interest to declare for this publication.

## ACKNOWLEDGEMENTS

The authors thank Graziano Bonelli for expert technical help.

## GLOSSARY

| | |
|---|---|
| **A1ATD** | α1-Antitrypsin Deficiency (OMIM #613490). |
| **ADA-SCID** | Adenosine deaminase deficiency-SCID (OMIM #103700). |
| **β-Thal** | β-thalassemia (OMIM #613985). |
| **CF** | Cystic Fibrosis disease (OMIM #219700). |
| **CRISPR/Cas (CRISPR associated) systems** | Clustered Regulatory Interspaced Short Palindromic Repeats or CRISPR are loci that contain multiple short direct repeats. CRISPR systems rely on crRNA and tracrRNA for sequence-specific silencing of invading foreign DNA. There are three types of CRISPR/Cas systems exist and in type II systems, Cas9 serves as an RNA-guided DNA endonuclease that cleaves DNA upon crRNA-tracrRNA target recognition. |
| **crRNA:** | CRISPR RNA base pair with tracrRNA to form a two-RNA structure that guides the Cas9 endonuclease to complementary DNA sites for cleavage. |
| **DMD** | Duchenne Muscular Dystrophy (OMIM #310200). |
| **DSB** | double-strand breaks are a form of DNA damage that occurs when both DNA strands are cleaved. |

**Gene editing**  is an approach in which the <u>genome</u> sequence is directly modified by insertions, substitutions or deletions of <u>DNA</u> bases through artificially engineered enzymes called nucleases, that use a sequence-specific DNA- binding domain and a non-specific DNA cleavage domain.

**HDR/HR**  Homology-directed repair is a template-dependent pathway for DSB repair providing the insertion of the targeted locus of single or multiple transgene.

**hESCs**  are pluripotent stem cells derived from the inner cell mass of embryos at the blastocyst stage.

**hiPSCs**  are induced pluripotent stem cells derived from adult somatic cell artificially reprogrammed.

**NETs:**  neutrophil extracellular traps composed of chromatin and granule proteins that are able to bind and kill extracellular microorganism.

**NHEJ**  Non-homologous end joining is a DSB repair pathway that ligates or joins two broken ends together. NHEJ does not use a homologous template for repair and thus can occur the introduction of small insertions and deletions at the site of the break, often inducing frameshifts that knockout gene function.

**PAM**  Proto-spacer adjacent motifs are short nucleotide motifs that occur on crRNA and are specifically recognized and required by Cas9 for DNA cleavage.

**SCD**  Sickle Cell Disease (OMIM #603903).

**SCID-X1**  X-linked Severe Combined Immunodeficiency (OMIM #300400).

**sgRNA**  single guide RNA.

**TALENs**  Transcription activator-like effector (TALE) nucleases are fusions of the FokI cleavage domain and DNA-binding domains derived from TALE proteins. TALEs contain multiple 33–35 amino acid repeat domains that each recognizes a single base pair. TALENs induce targeted DSBs that activate DNA damage response pathways and enable custom alterations.

**tracrRNA**  trans-activating chimeric RNA are non-coding RNA that promote crRNA processing and are required for activating RNA-guided cleavage by Cas9.

**WAS**  Wiskott-Aldrich syndrome (OMIM #301000).

**X-CGD**  X-Linked Chronic Granulomatous Disease (OMIM #300481).

**ZNF**  Zinc-finger nucleases are fusions of the non-specific DNA cleavage domain from the FokI restriction endonuclease with zinc-finger proteins. ZFN dimers induce targeted DNA double-strand breaks (DSBs) that stimulate DNA damage response pathways. The binding specificity of the designed zinc-finger domain directs the ZFN to a specific genomic site.

# REFERENCES

[1]   Thomson JA, Itskovitz-Eldor J, Shapiro SS, *et al.* Embryonic stem cell lines derived from human blastocysts. Science 1998; 282(5391): 1145-7.
[http://dx.doi.org/10.1126/science.282.5391.1145] [PMID: 9804556]

[2]   Wilmut I. Consternation and confusion following EU patent judgment. Cell Stem Cell 2011; 9(6): 498-9.
[http://dx.doi.org/10.1016/j.stem.2011.11.002] [PMID: 22136922]

[3]   Fortier LA. Stem cells: classifications, controversies, and clinical applications. Vet Surg 2005; 34(5): 415-23.
[http://dx.doi.org/10.1111/j.1532-950X.2005.00063.x] [PMID: 16266332]

[4] Phinney DG, Prockop DJ. Concise review: mesenchymal stem/multipotent stromal cells: the state of transdifferentiation and modes of tissue repair current views. Stem Cells 2007; 25(11): 2896-902.
[http://dx.doi.org/10.1634/stemcells.2007-0637] [PMID: 17901396]

[5] Alvarez CV, Garcia-Lavandeira M, Garcia-Rendueles ME, *et al.* Defining stem cell types: understanding the therapeutic potential of ESCs, ASCs, and iPS cells. J Mol Endocrinol 2012; 49(2): R89-R111.
[http://dx.doi.org/10.1530/JME-12-0072] [PMID: 22822049]

[6] English K, Wood KJ. Immunogenicity of embryonic stem cell-derived progenitors after transplantation. Curr Opin Organ Transplant 2011; 16(1): 90-5.
[http://dx.doi.org/10.1097/MOT.0b013e3283424faa] [PMID: 21150615]

[7] Takahashi K, Tanabe K, Ohnuki M, *et al.* Induction of pluripotent stem cells from adult human fibroblasts by defined factors. Cell 2007; 131(5): 861-72.
[http://dx.doi.org/10.1016/j.cell.2007.11.019] [PMID: 18035408]

[8] Bellin M, Marchetto MC, Gage FH, Mummery CL. Induced pluripotent stem cells: the new patient? Nat Rev Mol Cell Biol 2012; 13(11): 713-26.
[http://dx.doi.org/10.1038/nrm3448] [PMID: 23034453]

[9] Mallick KK, Cox SC. Biomaterial scaffolds for tissue engineering. Front Biosci (Elite Ed) 2013; 5: 341-60.
[http://dx.doi.org/10.2741/E620]

[10] Ilic D, Polak JM. Stem cells in regenerative medicine: introduction. Br Med Bull 2011; 98: 117-26.
[http://dx.doi.org/10.1093/bmb/ldr012] [PMID: 21565803]

[11] Webber DM, MacLeod SL, Bamshad MJ, *et al.* Developments in our understanding of the genetic basis of birth defects. Birth Defects Res A Clin Mol Teratol 2015; 103(8): 680-91.
[http://dx.doi.org/10.1002/bdra.23385] [PMID: 26033863]

[12] Wood AJ, Lo TW, Zeitler B, *et al.* Targeted genome editing across species using ZFNs and TALENs. Science 2011; 333(6040): 307.
[http://dx.doi.org/10.1126/science.1207773] [PMID: 21700836]

[13] Cermak T, Doyle EL, Christian M, *et al.* Efficient design and assembly of custom TALEN and other TAL effector-based constructs for DNA targeting. Nucleic Acids Res 2011; 39(12): e82.
[http://dx.doi.org/10.1093/nar/gkr218] [PMID: 21493687]

[14] Pennisi E. The CRISPR craze. Science 2013; 341(6148): 833-6.
[http://dx.doi.org/10.1126/science.341.6148.833] [PMID: 23970676]

[15] Ding Q, Regan SN, Xia Y, Oostrom LA, Cowan CA, Musunuru K. Enhanced efficiency of human pluripotent stem cell genome editing through replacing TALENs with CRISPRs. Cell Stem Cell 2013; 12(4): 393-4.
[http://dx.doi.org/10.1016/j.stem.2013.03.006] [PMID: 23561441]

[16] Zhang F, Wen Y, Guo X. CRISPR/Cas9 for genome editing: progress, implications and challenges. Hum Mol Genet 2014; 23(R1): R40-6.
[http://dx.doi.org/10.1093/hmg/ddu125] [PMID: 24651067]

[17]  Yu J, Vodyanik MA, Smuga-Otto K, *et al.* Induced pluripotent stem cell lines derived from human somatic cells. Science 2007; 318(5858): 1917-20.
[http://dx.doi.org/10.1126/science.1151526] [PMID: 18029452]

[18]  Papapetrou EP, Tomishima MJ, Chambers SM, *et al.* Stoichiometric and temporal requirements of Oct4, Sox2, Klf4, and c-Myc expression for efficient human iPSC induction and differentiation. Proc Natl Acad Sci USA 2009; 106(31): 12759-64.
[http://dx.doi.org/10.1073/pnas.0904825106] [PMID: 19549847]

[19]  Chen J, Liu J, Yang J, *et al.* BMPs functionally replace Klf4 and support efficient reprogramming of mouse fibroblasts by Oct4 alone. Cell Res 2011; 21(1): 205-12.
[http://dx.doi.org/10.1038/cr.2010.172] [PMID: 21135873]

[20]  Nakagawa M, Koyanagi M, Tanabe K, *et al.* Generation of induced pluripotent stem cells without Myc from mouse and human fibroblasts. Nat Biotechnol 2008; 26(1): 101-6.
[http://dx.doi.org/10.1038/nbt1374] [PMID: 18059259]

[21]  Wernig M, Meissner A, Cassady JP, Jaenisch R. c-Myc is dispensable for direct reprogramming of mouse fibroblasts. Cell Stem Cell 2008; 2(1): 10-2.
[http://dx.doi.org/10.1016/j.stem.2007.12.001] [PMID: 18371415]

[22]  Maekawa M, Yamaguchi K, Nakamura T, *et al.* Direct reprogramming of somatic cells is promoted by maternal transcription factor Glis1. Nature 2011; 474(7350): 225-9.
[http://dx.doi.org/10.1038/nature10106] [PMID: 21654807]

[23]  Nakagawa M, Takizawa N, Narita M, Ichisaka T, Yamanaka S. Promotion of direct reprogramming by transformation-deficient Myc. Proc Natl Acad Sci USA 2010; 107(32): 14152-7.
[http://dx.doi.org/10.1073/pnas.1009374107] [PMID: 20660764]

[24]  Hanna J, Cheng AW, Saha K, *et al.* Human embryonic stem cells with biological and epigenetic characteristics similar to those of mouse ESCs. Proc Natl Acad Sci USA 2010; 107(20): 9222-7.
[http://dx.doi.org/10.1073/pnas.1004584107] [PMID: 20442331]

[25]  Buecker C, Chen HH, Polo JM, *et al.* A murine ESC-like state facilitates transgenesis and homologous recombination in human pluripotent stem cells. Cell Stem Cell 2010; 6(6): 535-46.
[http://dx.doi.org/10.1016/j.stem.2010.05.003] [PMID: 20569691]

[26]  Li R, Liang J, Ni S, *et al.* A mesenchymal-to-epithelial transition initiates and is required for the nuclear reprogramming of mouse fibroblasts. Cell Stem Cell 2010; 7(1): 51-63.
[http://dx.doi.org/10.1016/j.stem.2010.04.014] [PMID: 20621050]

[27]  Ichida JK, Blanchard J, Lam K, *et al.* A small-molecule inhibitor of tgf-Beta signaling replaces sox2 in reprogramming by inducing nanog. Cell Stem Cell 2009; 5(5): 491-503.
[http://dx.doi.org/10.1016/j.stem.2009.09.012] [PMID: 19818703]

[28]  Stadtfeld M, Apostolou E, Ferrari F, *et al.* Ascorbic acid prevents loss of Dlk1-Dio3 imprinting and facilitates generation of all-iPS cell mice from terminally differentiated B cells. Nat Genet 2012; 44(4): 398-405, S1-S2.
[http://dx.doi.org/10.1038/ng.1110] [PMID: 22387999]

[29]  Takahashi K, Yamanaka S. Induction of pluripotent stem cells from mouse embryonic and adult fibroblast cultures by defined factors. Cell 2006; 126(4): 663-76.

[http://dx.doi.org/10.1016/j.cell.2006.07.024] [PMID: 16904174]

[30] González F, Boué S, Izpisúa Belmonte JC. Methods for making induced pluripotent stem cells: reprogramming à la carte. Nat Rev Genet 2011; 12(4): 231-42.
[http://dx.doi.org/10.1038/nrg2937] [PMID: 21339765]

[31] Carey BW, Markoulaki S, Hanna J, *et al.* Reprogramming of murine and human somatic cells using a single polycistronic vector. Proc Natl Acad Sci USA 2009; 106(1): 157-62.
[http://dx.doi.org/10.1073/pnas.0811426106] [PMID: 19109433]

[32] Sommer CA, Sommer AG, Longmire TA, *et al.* Excision of reprogramming transgenes improves the differentiation potential of iPS cells generated with a single excisable vector. Stem Cells 2010; 28(1): 64-74.
[PMID: 19904830]

[33] Kim D, Kim CH, Moon JI, *et al.* Generation of human induced pluripotent stem cells by direct delivery of reprogramming proteins. Cell Stem Cell 2009; 4(6): 472-6.
[http://dx.doi.org/10.1016/j.stem.2009.05.005] [PMID: 19481515]

[34] Ban H, Nishishita N, Fusaki N, *et al.* Efficient generation of transgene-free human induced pluripotent stem cells (iPSCs) by temperature-sensitive Sendai virus vectors. Proc Natl Acad Sci USA 2011; 108(34): 14234-9.
[http://dx.doi.org/10.1073/pnas.1103509108] [PMID: 21821793]

[35] Okita K, Matsumura Y, Sato Y, *et al.* A more efficient method to generate integration-free human iPS cells. Nat Methods 2011; 8(5): 409-12.
[http://dx.doi.org/10.1038/nmeth.1591] [PMID: 21460823]

[36] Warren L, Manos PD, Ahfeldt T, *et al.* Highly efficient reprogramming to pluripotency and directed differentiation of human cells with synthetic modified mRNA. Cell Stem Cell 2010; 7(5): 618-30.
[http://dx.doi.org/10.1016/j.stem.2010.08.012] [PMID: 20888316]

[37] Hou P, Li Y, Zhang X, *et al.* Pluripotent stem cells induced from mouse somatic cells by small-molecule compounds. Science 2013; 341(6146): 651-4.
[http://dx.doi.org/10.1126/science.1239278] [PMID: 23868920]

[38] Yakubov E, Rechavi G, Rozenblatt S, Givol D. Reprogramming of human fibroblasts to pluripotent stem cells using mRNA of four transcription factors. Biochem Biophys Res Commun 2010; 394(1): 189-93.
[http://dx.doi.org/10.1016/j.bbrc.2010.02.150] [PMID: 20188704]

[39] Rosa A, Brivanlou AH. Synthetic mRNAs: powerful tools for reprogramming and differentiation of human cells. Cell Stem Cell 2010; 7(5): 549-50.
[http://dx.doi.org/10.1016/j.stem.2010.10.002] [PMID: 21040893]

[40] Mikkelsen TS, Hanna J, Zhang X, *et al.* Dissecting direct reprogramming through integrative genomic analysis. Nature 2008; 454(7200): 49-55.
[http://dx.doi.org/10.1038/nature07056] [PMID: 18509334]

[41] Huangfu D, Osafune K, Maehr R, *et al.* Induction of pluripotent stem cells from primary human fibroblasts with only Oct4 and Sox2. Nat Biotechnol 2008; 26(11): 1269-75.
[http://dx.doi.org/10.1038/nbt.1502] [PMID: 18849973]

[42]    Esteban MA, Wang T, Qin B, *et al.* Vitamin C enhances the generation of mouse and human induced pluripotent stem cells. Cell Stem Cell 2010; 6(1): 71-9.
[http://dx.doi.org/10.1016/j.stem.2009.12.001] [PMID: 20036631]

[43]    Mali P, Chou BK, Yen J, *et al.* Butyrate greatly enhances derivation of human induced pluripotent stem cells by promoting epigenetic remodeling and the expression of pluripotency-associated genes. Stem Cells 2010; 28(4): 713-20.
[http://dx.doi.org/10.1002/stem.402] [PMID: 20201064]

[44]    Moriguchi H, Chung RT, Mihara M, Sato C. Generation of human induced pluripotent stem cells from liver progenitor cells by only small molecules. Hepatology 2010; 52(3): 1169.
[http://dx.doi.org/10.1002/hep.23851] [PMID: 20812362]

[45]    Chen T, Shen L, Yu J, *et al.* Rapamycin and other longevity-promoting compounds enhance the generation of mouse induced pluripotent stem cells. Aging Cell 2011; 10(5): 908-11.
[http://dx.doi.org/10.1111/j.1474-9726.2011.00722.x] [PMID: 21615676]

[46]    Wang Q, Xu X, Li J, *et al.* Lithium, an anti-psychotic drug, greatly enhances the generation of induced pluripotent stem cells. Cell Res 2011; 21(10): 1424-35.
[http://dx.doi.org/10.1038/cr.2011.108] [PMID: 21727907]

[47]    Spitalieri P, Talarico VR, Murdocca M, Novelli G, Sangiuolo F. Human induced pluripotent stem cells for monogenic disease modelling and therapy. World J Stem Cells 2016; 8(4): 118-35.
[http://dx.doi.org/10.4252/wjsc.v8.i4.118] [PMID: 27114745]

[48]    Cai S, Chan YS, Shum DK. Induced pluripotent stem cells and neurological disease models. Sheng Li Xue Bao 2014; 66(1): 55-66.
[PMID: 24553870]

[49]    Kim C. Disease modeling and cell based therapy with iPSC: future therapeutic option with fast and safe application. Blood Res 2014; 49(1): 7-14.
[http://dx.doi.org/10.5045/br.2014.49.1.7] [PMID: 24724061]

[50]    Urnov FD, Rebar EJ, Holmes MC, Zhang HS, Gregory PD. Genome editing with engineered zinc finger nucleases. Nat Rev Genet 2010; 11(9): 636-46.
[http://dx.doi.org/10.1038/nrg2842] [PMID: 20717154]

[51]    Carroll D. Genome engineering with zinc-finger nucleases. Genetics 2011; 188(4): 773-82.
[http://dx.doi.org/10.1534/genetics.111.131433] [PMID: 21828278]

[52]    Wyman C, Kanaar R. DNA double-strand break repair: alls well that ends well. Annu Rev Genet 2006; 40: 363-83.
[http://dx.doi.org/10.1146/annurev.genet.40.110405.090451] [PMID: 16895466]

[53]    Joung JK, Sander JD. TALENs: a widely applicable technology for targeted genome editing. Nat Rev Mol Cell Biol 2013; 14(1): 49-55.
[http://dx.doi.org/10.1038/nrm3486] [PMID: 23169466]

[54]    Mali P, Yang L, Esvelt KM, *et al.* RNA-guided human genome engineering *via* Cas9. Science 2013; 339(6121): 823-6.
[http://dx.doi.org/10.1126/science.1232033] [PMID: 23287722]

[55]     Byrne SM, Mali P, Church GM. Genome editing in human stem cells. Methods Enzymol 2014; 546: 119-38.
[http://dx.doi.org/10.1016/B978-0-12-801185-0.00006-4] [PMID: 25398338]

[56]     Moehle EA, Rock JM, Lee YL, *et al.* Targeted gene addition into a specified location in the human genome using designed zinc finger nucleases. Proc Natl Acad Sci USA 2007; 104(9): 3055-60.
[http://dx.doi.org/10.1073/pnas.0611478104] [PMID: 17360608]

[57]     Gaj T, Gersbach CA, Barbas CF III. ZFN, TALEN, and CRISPR/Cas-based methods for genome engineering. Trends Biotechnol 2013; 31(7): 397-405.
[http://dx.doi.org/10.1016/j.tibtech.2013.04.004] [PMID: 23664777]

[58]     Orlando SJ, Santiago Y, DeKelver RC, *et al.* Zinc-finger nuclease-driven targeted integration into mammalian genomes using donors with limited chromosomal homology. Nucleic Acids Res 2010; 38(15): e152.
[http://dx.doi.org/10.1093/nar/gkq512] [PMID: 20530528]

[59]     Chen F, Pruett-Miller SM, Huang Y, *et al.* High-frequency genome editing using ssDNA oligonucleotides with zinc-finger nucleases. Nat Methods 2011; 8(9): 753-5.
[http://dx.doi.org/10.1038/nmeth.1653] [PMID: 21765410]

[60]     Kim H, Kim JS. A guide to genome engineering with programmable nucleases. Nat Rev Genet 2014; 15(5): 321-34.
[http://dx.doi.org/10.1038/nrg3686] [PMID: 24690881]

[61]     Miller JC, Holmes MC, Wang J, *et al.* An improved zinc-finger nuclease architecture for highly specific genome editing. Nat Biotechnol 2007; 25(7): 778-85.
[http://dx.doi.org/10.1038/nbt1319] [PMID: 17603475]

[62]     Szczepek M, Brondani V, Büchel J, Serrano L, Segal DJ, Cathomen T. Structure-based redesign of the dimerization interface reduces the toxicity of zinc-finger nucleases. Nat Biotechnol 2007; 25(7): 786-93.
[http://dx.doi.org/10.1038/nbt1317] [PMID: 17603476]

[63]     Tupler R, Perini G, Green MR. Expressing the human genome. Nature 2001; 409(6822): 832-3.
[http://dx.doi.org/10.1038/35057011] [PMID: 11237001]

[64]     Wolfe SA, Nekludova L, Pabo CO. DNA recognition by Cys2His2 zinc finger proteins. Annu Rev Biophys Biomol Struct 2000; 29: 183-212.
[http://dx.doi.org/10.1146/annurev.biophys.29.1.183] [PMID: 10940247]

[65]     Rebar EJ, Pabo CO. Zinc finger phage: affinity selection of fingers with new DNA-binding specificities. Science 1994; 263(5147): 671-3.
[http://dx.doi.org/10.1126/science.8303274] [PMID: 8303274]

[66]     Bae KH, Kwon YD, Shin HC, *et al.* Human zinc fingers as building blocks in the construction of artificial transcription factors. Nat Biotechnol 2003; 21(3): 275-80.
[http://dx.doi.org/10.1038/nbt796] [PMID: 12592413]

[67]     Kim HJ, Lee HJ, Kim H, Cho SW, Kim JS. Targeted genome editing in human cells with zinc finger nucleases constructed via modular assembly. Genome Res 2009; 19(7): 1279-88.
[http://dx.doi.org/10.1101/gr.089417.108] [PMID: 19470664]

[68]   Sebastiano V, Maeder ML, Angstman JF, *et al. In situ* genetic correction of the sickle cell anemia mutation in human induced pluripotent stem cells using engineered zinc finger nucleases. Stem Cells 2011; 29(11): 1717-26.
[http://dx.doi.org/10.1002/stem.718] [PMID: 21898685]

[69]   Zou J, Mali P, Huang X, Dowey SN, Cheng L. Site-specific gene correction of a point mutation in human iPS cells derived from an adult patient with sickle cell disease. Blood 2011; 118(17): 4599-608.
[http://dx.doi.org/10.1182/blood-2011-02-335554] [PMID: 21881051]

[70]   Weatherall DJ, Clegg JB. Inherited haemoglobin disorders: an increasing global health problem. Bull World Health Organ 2001; 79(8): 704-12.
[PMID: 11545326]

[71]   Maeder ML, Thibodeau-Beganny S, Sander JD, Voytas DF, Joung JK. Oligomerized pool engineering (OPEN): an open-source protocol for making customized zinc-finger arrays. Nat Protoc 2009; 4(10): 1471-501.
[http://dx.doi.org/10.1038/nprot.2009.98] [PMID: 19798082]

[72]   Chang CJ, Bouhassira EE. Zinc-finger nuclease-mediated correction of α-thalassemia in iPS cells. Blood 2012; 120(19): 3906-14.
[http://dx.doi.org/10.1182/blood-2012-03-420703] [PMID: 23002118]

[73]   Soldner F, Laganière J, Cheng AW, *et al.* Generation of isogenic pluripotent stem cells differing exclusively at two early onset Parkinson point mutations. Cell 2011; 146(2): 318-31.
[http://dx.doi.org/10.1016/j.cell.2011.06.019] [PMID: 21757228]

[74]   Kerem B, Rommens JM, Buchanan JA, *et al.* Identification of the cystic fibrosis gene: genetic analysis. Science 1989; 245(4922): 1073-80.
[http://dx.doi.org/10.1126/science.2570460] [PMID: 2570460]

[75]   Crane AM, Kramer P, Bui JH, *et al.* Targeted correction and restored function of the CFTR gene in cystic fibrosis induced pluripotent stem cells. Stem Cell Rep 2015; 4(4): 569-77.
[http://dx.doi.org/10.1016/j.stemcr.2015.02.005] [PMID: 25772471]

[76]   Antonarakis SE, Lyle R, Dermitzakis ET, Reymond A, Deutsch S. Chromosome 21 and down syndrome: from genomics to pathophysiology. Nat Rev Genet 2004; 5(10): 725-38.
[http://dx.doi.org/10.1038/nrg1448] [PMID: 15510164]

[77]   Lott IT, Dierssen M. Cognitive deficits and associated neurological complications in individuals with Downs syndrome. Lancet Neurol 2010; 9(6): 623-33.
[http://dx.doi.org/10.1016/S1474-4422(10)70112-5] [PMID: 20494326]

[78]   Haydar TF, Reeves RH. Trisomy 21 and early brain development. Trends Neurosci 2012; 35(2): 81-91.
[http://dx.doi.org/10.1016/j.tins.2011.11.001] [PMID: 22169531]

[79]   Guidi S, Ciani E, Bonasoni P, Santini D, Bartesaghi R. Widespread proliferation impairment and hypocellularity in the cerebellum of fetuses with down syndrome. Brain Pathol 2011; 21(4): 361-73.
[http://dx.doi.org/10.1111/j.1750-3639.2010.00459.x] [PMID: 21040072]

[80]   Jiang J, Jing Y, Cost GJ, *et al.* Translating dosage compensation to trisomy 21. Nature 2013; 500(7462): 296-300.

[http://dx.doi.org/10.1038/nature12394] [PMID: 23863942]

[81] Perlmutter DH. Autophagic disposal of the aggregation-prone protein that causes liver inflammation and carcinogenesis in alpha-1-antitrypsin deficiency. Cell Death Differ 2009; 16(1): 39-45.
[http://dx.doi.org/10.1038/cdd.2008.103] [PMID: 18617899]

[82] Gooptu B, Lomas DA. Conformational pathology of the serpins: themes, variations, and therapeutic strategies. Annu Rev Biochem 2009; 78: 147-76.
[http://dx.doi.org/10.1146/annurev.biochem.78.082107.133320] [PMID: 19245336]

[83] Yusa K, Rashid ST, Strick-Marchand H, *et al.* Targeted gene correction of α1-antitrypsin deficiency in induced pluripotent stem cells. Nature 2011; 478(7369): 391-4.
[http://dx.doi.org/10.1038/nature10424] [PMID: 21993621]

[84] Winkelstein JA, Marino MC, Johnston RB Jr, *et al.* Chronic granulomatous disease. Report on a national registry of 368 patients. Medicine (Baltimore) 2000; 79(3): 155-69.
[http://dx.doi.org/10.1097/00005792-200005000-00003] [PMID: 10844935]

[85] van den Berg JM, van Koppen E, Åhlin A, *et al.* Chronic granulomatous disease: the European experience. PLoS One 2009; 4(4): e5234.
[http://dx.doi.org/10.1371/journal.pone.0005234] [PMID: 19381301]

[86] Zou J, Sweeney CL, Chou BK, *et al.* Oxidase-deficient neutrophils from X-linked chronic granulomatous disease iPS cells: functional correction by zinc finger nuclease-mediated safe harbor targeting. Blood 2011; 117(21): 5561-72.
[http://dx.doi.org/10.1182/blood-2010-12-328161] [PMID: 21411759]

[87] Boch J, Scholze H, Schornack S, *et al.* Breaking the code of DNA binding specificity of TAL-type III effectors. Science 2009; 326(5959): 1509-12.
[http://dx.doi.org/10.1126/science.1178811] [PMID: 19933107]

[88] Moscou MJ, Bogdanove AJ. A simple cipher governs DNA recognition by TAL effectors. Science 2009; 326(5959): 1501.
[http://dx.doi.org/10.1126/science.1178817] [PMID: 19933106]

[89] Mak AN, Bradley P, Cernadas RA, Bogdanove AJ, Stoddard BL. The crystal structure of TAL effector PthXo1 bound to its DNA target. Science 2012; 335(6069): 716-9.
[http://dx.doi.org/10.1126/science.1216211] [PMID: 22223736]

[90] Deng D, Yan C, Pan X, *et al.* Structural basis for sequence-specific recognition of DNA by TAL effectors. Science 2012; 335(6069): 720-3.
[http://dx.doi.org/10.1126/science.1215670] [PMID: 22223738]

[91] Reyon D, Tsai SQ, Khayter C, Foden JA, Sander JD, Joung JK. FLASH assembly of TALENs for high-throughput genome editing. Nat Biotechnol 2012; 30(5): 460-5.
[http://dx.doi.org/10.1038/nbt.2170] [PMID: 22484455]

[92] Schmid-Burgk JL, Schmidt T, Kaiser V, Höning K, Hornung V. A ligation-independent cloning technique for high-throughput assembly of transcription activator–like effector genes. Nat Biotechnol 2013; 31(1): 76-81.
[http://dx.doi.org/10.1038/nbt.2460] [PMID: 23242165]

[93] Kim Y, Kweon J, Kim A, *et al.* A library of TAL effector nucleases spanning the human genome. Nat

Biotechnol 2013; 31(3): 251-8.
[http://dx.doi.org/10.1038/nbt.2517] [PMID: 23417094]

[94]   Hockemeyer D, Wang H, Kiani S, *et al.* Genetic engineering of human pluripotent cells using TALE nucleases. Nat Biotechnol 2011; 29(8): 731-4.
[http://dx.doi.org/10.1038/nbt.1927] [PMID: 21738127]

[95]   Hockemeyer D, Wang H, Kiani S, *et al.* Genetic engineering of human pluripotent cells using TALE nucleases. Nat Biotechnol 2011; 29(8): 731-4.
[http://dx.doi.org/10.1038/nbt.1927] [PMID: 21738127]

[96]   Ding Q, Lee YK, Schaefer EA, *et al.* A TALEN genome-editing system for generating human stem cell-based disease models. Cell Stem Cell 2013; 12(2): 238-51.
[http://dx.doi.org/10.1016/j.stem.2012.11.011] [PMID: 23246482]

[97]   Dreyer AK, Hoffmann D, Lachmann N, *et al.* TALEN-mediated functional correction of X-linked chronic granulomatous disease in patient-derived induced pluripotent stem cells. Biomaterials 2015; 69: 191-200.
[http://dx.doi.org/10.1016/j.biomaterials.2015.07.057] [PMID: 26295532]

[98]   Pattanayak V, Ramirez CL, Joung JK, Liu DR. Revealing off-target cleavage specificities of zinc-finger nucleases by in vitro selection. Nat Methods 2011; 8(9): 765-70.
[http://dx.doi.org/10.1038/nmeth.1670] [PMID: 21822273]

[99]   Sun N, Zhao H. Seamless correction of the sickle cell disease mutation of the HBB gene in human induced pluripotent stem cells using TALENs. Biotechnol Bioeng 2014; 111(5): 1048-53.
[http://dx.doi.org/10.1002/bit.25018] [PMID: 23928856]

[100]  Choi SM, Kim Y, Shim JS, *et al.* Efficient drug screening and gene correction for treating liver disease using patient-specific stem cells. Hepatology 2013; 57(6): 2458-68.
[http://dx.doi.org/10.1002/hep.26237] [PMID: 23325555]

[101]  Osborn MJ, Starker CG, McElroy AN, *et al.* TALEN-based gene correction for epidermolysis bullosa. Mol Ther 2013; 21(6): 1151-9.
[http://dx.doi.org/10.1038/mt.2013.56] [PMID: 23546300]

[102]  Cao A, Galanello R. Beta-thalassemia. Genet Med 2010; 12(2): 61-76.
[http://dx.doi.org/10.1097/GIM.0b013e3181cd68ed] [PMID: 20098328]

[103]  Ma N, Liao B, Zhang H, *et al.* Transcription activator-like effector nuclease (TALEN)-mediated gene correction in integration-free β-thalassemia induced pluripotent stem cells. J Biol Chem 2013; 288(48): 34671-9.
[http://dx.doi.org/10.1074/jbc.M113.496174] [PMID: 24155235]

[104]  Carstea ED, Morris JA, Coleman KG, *et al.* Niemann-Pick C1 disease gene: homology to mediators of cholesterol homeostasis. Science 1997; 277(5323): 228-31.
[http://dx.doi.org/10.1126/science.277.5323.228] [PMID: 9211849]

[105]  Vanier MT. Niemann-Pick disease type C. Orphanet J Rare Dis 2010; 5: 16.
[http://dx.doi.org/10.1186/1750-1172-5-16] [PMID: 20525256]

[106]  Maetzel D, Sarkar S, Wang H, *et al.* Genetic and chemical correction of cholesterol accumulation and impaired autophagy in hepatic and neural cells derived from Niemann-Pick Type C patient-specific

iPS cells. Stem Cell Rep 2014; 2(6): 866-80.
[http://dx.doi.org/10.1016/j.stemcr.2014.03.014] [PMID: 24936472]

[107]   Ishino Y, Shinagawa H, Makino K, Amemura M, Nakata A. Nucleotide sequence of the iap gene, responsible for alkaline phosphatase isozyme conversion in *Escherichia coli*, and identification of the gene product. J Bacteriol 1987; 169(12): 5429-33.
[http://dx.doi.org/10.1128/jb.169.12.5429-5433.1987] [PMID: 3316184]

[108]   Mojica FJ, Díez-Villaseñor C, Soria E, Juez G. Biological significance of a family of regularly spaced repeats in the genomes of Archaea, Bacteria and mitochondria. Mol Microbiol 2000; 36(1): 244-6.
[http://dx.doi.org/10.1046/j.1365-2958.2000.01838.x] [PMID: 10760181]

[109]   Bolotin A, Quinquis B, Sorokin A, Ehrlich SD. Clustered regularly interspaced short palindrome repeats (CRISPRs) have spacers of extrachromosomal origin. Microbiology 2005; 151(Pt 8): 2551-61.
[http://dx.doi.org/10.1099/mic.0.28048-0] [PMID: 16079334]

[110]   Mojica FJ, Díez-Villaseñor C, García-Martínez J, Soria E. Intervening sequences of regularly spaced prokaryotic repeats derive from foreign genetic elements. J Mol Evol 2005; 60(2): 174-82.
[http://dx.doi.org/10.1007/s00239-004-0046-3] [PMID: 15791728]

[111]   Pourcel C, Salvignol G, Vergnaud G. CRISPR elements in Yersinia pestis acquire new repeats by preferential uptake of bacteriophage DNA, and provide additional tools for evolutionary studies. Microbiology 2005; 151(Pt 3): 653-63.
[http://dx.doi.org/10.1099/mic.0.27437-0] [PMID: 15758212]

[112]   Tang TH, Bachellerie JP, Rozhdestvensky T, *et al*. Identification of 86 candidates for small non-messenger RNAs from the archaeon Archaeoglobus fulgidus. Proc Natl Acad Sci USA 2002; 99(11): 7536-41.
[http://dx.doi.org/10.1073/pnas.112047299] [PMID: 12032318]

[113]   Jansen R, Embden JD, Gaastra W, Schouls LM. Identification of genes that are associated with DNA repeats in prokaryotes. Mol Microbiol 2002; 43(6): 1565-75.
[http://dx.doi.org/10.1046/j.1365-2958.2002.02839.x] [PMID: 11952905]

[114]   Haft DH, Selengut J, Mongodin EF, Nelson KE. A guild of 45 CRISPR-associated (Cas) protein families and multiple CRISPR/Cas subtypes exist in prokaryotic genomes. PLOS Comput Biol 2005; 1(6): e60.
[http://dx.doi.org/10.1371/journal.pcbi.0010060] [PMID: 16292354]

[115]   Makarova KS, Grishin NV, Shabalina SA, Wolf YI, Koonin EV. A putative RNA-interference-based immune system in prokaryotes: computational analysis of the predicted enzymatic machinery, functional analogies with eukaryotic RNAi, and hypothetical mechanisms of action. Biol Direct 2006; 1: 7.
[http://dx.doi.org/10.1186/1745-6150-1-7] [PMID: 16545108]

[116]   Barrangou R, Fremaux C, Deveau H, *et al*. CRISPR provides acquired resistance against viruses in prokaryotes. Science 2007; 315(5819): 1709-12.
[http://dx.doi.org/10.1126/science.1138140] [PMID: 17379808]

[117]   Cong L, Ran FA, Cox D, *et al*. Multiplex genome engineering using CRISPR/Cas systems. Science 2013; 339(6121): 819-23.
[http://dx.doi.org/10.1126/science.1231143] [PMID: 23287718]

[118]  Deltcheva E, Chylinski K, Sharma CM, *et al.* CRISPR RNA maturation by trans-encoded small RNA and host factor RNase III. Nature 2011; 471(7340): 602-7.
[http://dx.doi.org/10.1038/nature09886] [PMID: 21455174]

[119]  Jinek M, Chylinski K, Fonfara I, Hauer M, Doudna JA, Charpentier E. A programmable dual-RN--guided DNA endonuclease in adaptive bacterial immunity. Science 2012; 337(6096): 816-21.
[http://dx.doi.org/10.1126/science.1225829] [PMID: 22745249]

[120]  Makarova KS, Haft DH, Barrangou R, *et al.* Evolution and classification of the CRISPR-Cas systems. Nat Rev Microbiol 2011; 9(6): 467-77.
[http://dx.doi.org/10.1038/nrmicro2577] [PMID: 21552286]

[121]  Nam KH, Haitjema C, Liu X, *et al.* Cas5d protein processes pre-crRNA and assembles into a cascade-like interference complex in subtype I-C/Dvulg CRISPR-Cas system. Structure 2012; 20(9): 1574-84.
[http://dx.doi.org/10.1016/j.str.2012.06.016] [PMID: 22841292]

[122]  Haurwitz RE, Jinek M, Wiedenheft B, Zhou K, Doudna JA. Sequence- and structure-specific RNA processing by a CRISPR endonuclease. Science 2010; 329(5997): 1355-8.
[http://dx.doi.org/10.1126/science.1192272] [PMID: 20829488]

[123]  Hatoum-Aslan A, Maniv I, Marraffini LA. Mature clustered, regularly interspaced, short palindromic repeats RNA (crRNA) length is measured by a ruler mechanism anchored at the precursor processing site. Proc Natl Acad Sci USA 2011; 108(52): 21218-22.
[http://dx.doi.org/10.1073/pnas.1112832108] [PMID: 22160698]

[124]  Doudna JA, Charpentier E. Genome editing. The new frontier of genome engineering with CRISPR-Cas9. Science 2014; 346(6213): 1258096.
[http://dx.doi.org/10.1126/science.1258096] [PMID: 25430774]

[125]  Ding Q, Regan SN, Xia Y, Oostrom LA, Cowan CA, Musunuru K. Enhanced efficiency of human pluripotent stem cell genome editing through replacing TALENs with CRISPRs. Cell Stem Cell 2013; 12(4): 393-4.
[http://dx.doi.org/10.1016/j.stem.2013.03.006] [PMID: 23561441]

[126]  Fu Y, Foden JA, Khayter C, *et al.* High-frequency off-target mutagenesis induced by CRISPR-Cas nucleases in human cells. Nat Biotechnol 2013; 31(9): 822-6.
[http://dx.doi.org/10.1038/nbt.2623] [PMID: 23792628]

[127]  Garneau JE, Dupuis MÈ, Villion M, *et al.* The CRISPR/Cas bacterial immune system cleaves bacteriophage and plasmid DNA. Nature 2010; 468(7320): 67-71.
[http://dx.doi.org/10.1038/nature09523] [PMID: 21048762]

[128]  Chylinski K, Makarova KS, Charpentier E, Koonin EV. Classification and evolution of type II CRISPR-Cas systems. Nucleic Acids Res 2014; 42(10): 6091-105.
[http://dx.doi.org/10.1093/nar/gku241] [PMID: 24728998]

[129]  Hsu PD, Lander ES, Zhang F. Development and applications of CRISPR-Cas9 for genome engineering. Cell 2014; 157(6): 1262-78.
[http://dx.doi.org/10.1016/j.cell.2014.05.010] [PMID: 24906146]

[130]  Park KS, Lee DK, Lee H, *et al.* Phenotypic alteration of eukaryotic cells using randomized libraries of artificial transcription factors. Nat Biotechnol 2003; 21(10): 1208-14.

[http://dx.doi.org/10.1038/nbt868] [PMID: 12960965]

[131] Hu J, Lei Y, Wong WK, *et al.* Direct activation of human and mouse Oct4 genes using engineered TALE and Cas9 transcription factors. Nucleic Acids Res 2014; 42(7): 4375-90.
[http://dx.doi.org/10.1093/nar/gku109] [PMID: 24500196]

[132] Kabadi AM, Gersbach CA. Engineering synthetic TALE and CRISPR/Cas9 transcription factors for regulating gene expression. Methods 2014; 69(2): 188-97.
[http://dx.doi.org/10.1016/j.ymeth.2014.06.014] [PMID: 25010559]

[133] Li HL, Fujimoto N, Sasakawa N, *et al.* Precise correction of the dystrophin gene in duchenne muscular dystrophy patient induced pluripotent stem cells by TALEN and CRISPR-Cas9. Stem Cell Rep 2015; 4(1): 143-54.
[http://dx.doi.org/10.1016/j.stemcr.2014.10.013] [PMID: 25434822]

[134] Mercuri E, Muntoni F. Muscular dystrophy: new challenges and review of the current clinical trials. Curr Opin Pediatr 2013; 25(6): 701-7.
[http://dx.doi.org/10.1097/MOP.0b013e328365ace5] [PMID: 24240289]

[135] Braun R, Wang Z, Mack DL, Childers MK. Gene therapy for inherited muscle diseases: where genetics meets rehabilitation medicine. Am J Phys Med Rehabil 2014; 93(11) (Suppl. 3): S97-S107.
[http://dx.doi.org/10.1097/PHM.0000000000000138] [PMID: 25313664]

[136] Firth AL, Menon T, Parker GS, *et al.* Functional Gene Correction for Cystic Fibrosis in Lung Epithelial Cells Generated from Patient iPSCs. Cell Reports 2015; 12(9): 1385-90.
[http://dx.doi.org/10.1016/j.celrep.2015.07.062] [PMID: 26299960]

[137] Flynn R, Grundmann A, Renz P, *et al.* CRISPR-mediated genotypic and phenotypic correction of a chronic granulomatous disease mutation in human iPS cells. Exp Hematol 2015; 43(10): 838-848.e3.

[138] Chang CW, Lai YS, Westin E, *et al.* Modeling human severe combined immunodeficiency and correction by CRISPR/Cas9-enhanced gene targeting. Cell Reports 2015; 12(10): 1668-77.
[http://dx.doi.org/10.1016/j.celrep.2015.08.013] [PMID: 26321643]

[139] Park CY, Kim DH, Son JS, *et al.* Functional correction of large factor VIII gene chromosomal inversions in hemophilia a patient-derived iPSCs using CRISPR-Cas9. Cell Stem Cell 2015; 17(2): 213-20.
[http://dx.doi.org/10.1016/j.stem.2015.07.001] [PMID: 26212079]

[140] Chakraborty C, Shah KD, Cao WG, Hsu CH, Wen ZH, Lin CS. Potentialities of induced pluripotent stem (iPS) cells for treatment of diseases. Curr Mol Med 2010; 10(8): 756-62.
[http://dx.doi.org/10.2174/156652410793384178] [PMID: 20937020]

[141] Shtrichman R, Germanguz I, Itskovitz-Eldor J. Induced pluripotent stem cells (iPSCs) derived from different cell sources and their potential for regenerative and personalized medicine. Curr Mol Med 2013; 13(5): 792-805.
[http://dx.doi.org/10.2174/1566524011313050010] [PMID: 23642060]

[142] Park KM, Cha SH, Ahn C, Woo HM. Generation of porcine induced pluripotent stem cells and evaluation of their major histocompatibility complex protein expression *in vitro*. Vet Res Commun 2013; 37(4): 293-301.
[http://dx.doi.org/10.1007/s11259-013-9574-x] [PMID: 23975685]

[143]  Spitalieri P, Talarico RV, Botta A, *et al.* Generation of Human Induced Pluripotent Stem Cells from Extraembryonic Tissues of Fetuses Affected by Monogenic Diseases. Cell Reprogram 2015; 17(4): 275-87.
[http://dx.doi.org/10.1089/cell.2015.0003] [PMID: 26474030]

# CHAPTER 7

# Stem Cells for Therapeutic Delivery of Mediators and Drugs

**Pranela Rameshwar**[1,*], **Jimmy Patel**[1] and **Alexander Aleynik**[2]

*[1] New Jersey Medical School, Departmant of Medicine, Hematology/Oncology, New Jersey Medical School, NJ, USA*

*[2] Graduate School of Biomedical Health Sciences, Rutgers University, Newark, NJ, USA*

**Abstract:** This chapter discusses alternative methods for drug delivery. Specifically, we focused on the use of mesenchymal stem cells (MSCs) and to show how these stem cells can be available as off-the-shelf cells. The advantages of MSCs are their unique immune properties and the ability of these cells to migrate to areas of inflammation such as tumors. In addition, MSCs are easy to harvest with several fold expansion, as well as little to no ethical concern. Although MSCs are similar by phenotype, the effectiveness from each source needs to be compared for homing to the desired organ/tissue, intercellular communication and the delivery of non-coding RNA through secreted exosomes. A major advantage of MSCs is the ease by which they can become available as off-the-shelf cells containing the drugs or RNA for immediate transplant to patients. There is little concern that MSCs will linger for a prolonged period because the clinical and experimental evidence indicate that allogeneic MSCs (off-the-shelf) can be readily cleared by the immune system. The chapter discusses why there is an immediate need for cellular delivery of drugs, given the cumbersome regulation and confounds of current single drug trials.

**Keywords:** Breast cancer, Cancer stem cell, Connexin, Cytokines, Drug delivery, Exosomes, Mesenchymal stem cells, miRNA, Non-coding RNA.

---

* **Corresponding author Pranela Rameshwar:** Department of Medicine – Division of Hematology/Oncology, New Jersey Medical School, Rutgers School of Biomedical Health Science, Newark, NJ 07103 USA; Tel: (973) 972 0625; E-mail: rameshwa@njms.rutgers.edu

# INTRODUCTION

The current method to select a drug to treat diseases is generally based on a drug that has undergone costly clinical trials in systematic phases. The trial is accompanied by an overwhelming large package of the application to comply with the respective regulatory bodies. These trials are generally performed with clear inclusion and exclusion criteria for specific populations. This clearly indicates that the clinical trials have begun with a bias and, this bias could continue to marketing when those exclusion criteria may not be relevant.

The United States of America (USA), which has a diverse patient population with respect to ethnicity, is quite clear that all ethnic groups and sexes should be included in the clinical trials. This message is adhered by institutional regulatory boards that oversee the trials, with careful review in the justification for the proposed inclusion and exclusion criteria. Due to these regulations and the availability of diverse ethnic groups in the USA, the system, although not perfect, the clinical trials in the USA are generally inclusive with respect to ethnicity.

The importance of including individuals with different ethnic backgrounds seems to be overlooked as the cost of clinical trials increase, leading to the drug developers seeking countries that could conduct the trials at less cost. The question is not the integrity of the trial but the ability to conduct the trials at reduced cost, although at the expense of ethnic diversity. This is usually not deliberate but to conduct cheaper clinical trials in counties in which the general population is ethnically homogeneous. Despite these shortcomings, the drugs are eventually approved and used for all ethnic backgrounds.

Another issue currently in medicine is the over-use of the term 'precision medicine'. This is generally considered as the development of treatments to specifically suit the patient, despite similar diagnosis. It is assumed that this type of treatment will consider the varied polymorphisms in human but it is not clear if a large number of parameters are, or will be considered when the treatments are designed. Basic biologists and clinical scientists in general, are not open to include mathematicians and engineers in their teams. These latter individuals can develop simulation models that can be tested by biologists. This type of 'tunnel vision' of those involved in clinical trials has led to a relatively small window of soaring profits for the makers of the drugs. However, this could results in questionable long-term efficacy for the patients.

To reiterate the potential harm by excluding ethnic diversity in a drug trial pertains to the many failures of the drug. At this time, it is left to the individual physicians to fill the gap. This is rarely done since a controlled trial is costly. In some cases there are cooperative bodies that share the data from different patients within a country and with international bodies. However, it should be noted that the cost needed to fix the problem that was created early has to be paid by someone, and in most cases, by higher cost of the same drugs or others.

The field of stem cells is evolving as a different method of drug delivery. This article will use a model of breast cancer (BC) to discuss how stem cells can be developed for drug delivery. The premise is to understand how the microenvironment facilitates the disease, in this case, cancer dormancy or metastasis. The information will then identify drug targets. The targets will be blocked by drugs or small non-coding RNA (ncRNA) within stem cells. This method might eliminate the broader issue of ethnic diversity but would not eliminate the problem.

This article focuses on mesenchymal stem cells (MSCs), mostly due to the ease in obtaining large amounts, reduced ethical issues, and the unique ability of MSCs to be used given across allogeneic barrier [1]. We will also discuss the types of communications that could occur that will allow the MSCs to transfer the drugs. We will focus on RNA delivery since this is a growing field in therapeutics. The data has identified MSCs as good cellular vehicle to deliver small ncRNA or their antagomiRs.

**Mesenchymal Stem Cells (MSCs) - Background**

Since this review is focused on MSCs, we will present a brief background on these stem cells. This section will also include the advantages of MSCs, in particular, their use across allogeneic barrier, generally referred as `off-the-shelf' stem cells. MSCs are ubiquitously present in adults and in fetal tissues. In adults, MSCs are predominantly found in the bone marrow and adipose tissues. MSCs can be easily isolated from several tissues, including amniotic fluid, Wharton jelly of the umbilical cord and placenta [2]. Regardless of the sources, MSCs showed similarity with respect to phenotype and functionally, differentiating along multiple lineages.

Presently, it is unclear if MSCs from all sources are similar with regards to their ability to home efficiently to specific organs. For example, it cannot be stated that MSCs from different sources can home to the lungs or brain with similar efficiency. Answers to such questions are required if MSCs are selected as the vehicular stem cells for drug delivery.

MSCs show mesodermal properties despite evidence that they could originate in the neuroectoderm [3]. Similar to other nucleated cells, MSCs express the class I major histocompatibity complex (MHC-I) and a small subset, the MHC-II [4, 5]. As third party cells, when MSCs are placed within two different sources of immune cells, they can be licensed to become immune suppressor cells [6]. It is this property of MSCs that has led to their use in graft *vs.* host disease (GvHD) as third party cells with allogeneic donors cells for the transplant [7, 8]. The use of MSCs for inflammatory disorders is still considered early. Thus, the approval for MSCs has been given for steroid-resistant GvHD [9].

Although MSCs are thought as cells with immune suppressive properties, it should be noted that this function is not inherent to MSCs. In fact, MSCs need to be licensed, and this can occur when the MSCs are placed or homed to regions of inflammation, as would occur in GvHD [7]. This licensing ability of MSCs is similar to situations around and within tumors that are considered as regions of inflammation with high amounts of cytokines and extracellular matrices. Tumors therefore provide MSCs with a microenvironment that allows for their licensing as immune suppressors. At the same time, the MSCs would provide the tumors with a survival advantage due to the immune suppressive effects of MSCs.

The important issue is if MSCs could be used as a vehicle for direct delivery of the intended drugs to the targeted site, such metastatic sites of tumors. There are two broad issues in answers to this question. The first, which is the more straightforward explanation, is to determine whether MSCs can home to the site of tissue insult such as migrate to a metastatic or dormant region of the tumor. At this region, it is expected that the gradient of chemotactic factors to be increased. MSCs express several receptors for cytokines, including those that can recognize patterning antigen such as allergens [10]. The second, to determine if the MSCs, after homing to the desired region, can efficiently release the drug.

## Advantages/Disadvantages of MSCs in Drug Delivery for Cancer

The ability of MSCs to home to regions of tumors could be an advantage for drug delivery [11]. However, there are obvious disadvantage when MSCs are within or close to the tumors. MSCs interact with the tumor initiating cancer stem cells (CSCs) to protect them from immune clearance and to support cancer growth [12]. In the presence of BCCs, MSCs suppress the innate cytotoxic natural killer cells and at the same time, increase the immune suppressive regulatory T-cells, with the likelihood of enhancing the molecules involved in immune checkpoint [13 - 15]. Together, a tumor milieu, although attractive to MSCs, provides an advantage to the tumor. Thus, going forward, the challenge is to be able to use MSCs to deliver drugs without compromising the 'danger' of these stem cells at the tumor site.

As discussed above, MSCs can be used as 'off-the-shelf' cells, making them available on-demand for patients. This is an excellent advantage for these stem cells as vehicles of drug delivery. One can envision that the MSCs will be engineered with the drugs, thereby making them readily available on demand. This is in contrast to autologous MSCs that will need to be harvested and expanded.

There are two major disadvantages in the use of autologous MSCs. Firstly, to reiterate, it takes a long time to expand the harvested MSCs. More importantly, if the treatment is for BC, there could be problems to get adipose tissue for MSCs since the patient might be cachexia. In addition, there may be an ethical issue to perform an invasive procedure with a bone marrow aspirate to obtain MSCs.

The argument for allogeneic MSCs as a drug delivery is compelling, mostly due to their ability to evade immediate immune clearance. An added advantage is for a slow, but delayed ability of the immune system to clear the allogeneic MSCs [16].

The eventual clearance of MSCs is important since this will avoid the retention of MSCs from a different donor. The lingering allogeneic MSCs can cause a low level of immune response and can also interact with cells and soluble factors within a microenvironment to differentiate into unwanted cells [4, 17, 18]. As examples, the importance of eliminating the allogeneic MSCs after drug delivery will not cause MSC-derived cardiac cells to be established in the brain or the formation of bone within the cardiac tissue.

Studies have shown that even with differentiated cells, the MHC-II molecules can be re-expressed in the presence of interferon γ [5]. There is no strong report to show that allogeneic MSCs can integrate into tissues and develop into specialized cells [19, 20]. Nonetheless, if MSCs, left over from drug delivery, differentiate into few specialized cells, these differentiated cells may be sufficient to become immune stimulators. This could occur during an infection that will trigger an inflammation response. This would result in an increase in interferon γ that may induce the expression of MHC-II [5]. Due to the importance of this potential confound, we expand on this point. If there are residual allogeneic cells in the liver, there is no likelihood of an overt immune response. However, if there is viral infection, this could lead to systemic inflammation, resulting in the expression of allo-MHC-II. Since the residual cells within the liver can then become stimulations, this could lead to tissue damage.

There are proposals to engineer the MSCs with suicidal genes such as thymidine kinase and then add ganciclovir to eliminate the engineer cells after they have elicited the intended function [21, 22]. The principle of this method is straightforward. There is a need however, for robust studies to test the potential toxicity of such treatment. In the case of late stage cancer and those with poor prognosis such as glioblastomas, the cytotoxic genes could be among the first to be tested [11]. However, since MSC therapy is being considered for treating non-malignant disorders such as neurodegenerative diseases, the direct homing of MSCs to the brain has not been fully established [11, 23, 24]. More importantly, there are questions on the source of stem cells for delivering drugs to the brain. Neural stem cells (NSCs) are in clinical trials as cellular vehicle to deliver drugs for glioblastoma multiform [25]. As discussed above, it would be difficult to use autologous NSCs, underscoring the potential confounds with allogeneic NSCs.

This section cannot be completed without a statement about induced pluripotent stem cells (iPS) and embryonic stem cells (ESCs). There is no doubt that these two immature stem cells should be able to produce any type of specialized cells. However, a significant issue with the use of iPSCs and ESCs is the ease by which these cells transform [26]. This, however, does not imply that iPS and ESCs do not have a place in the field of cellular delivery of drugs. In fact, a recent report indicated that MSCs, derived from iPS cells, can home with better efficiency to tumors as compared to those from bone marrow and, with reduced ability to promote tumor growth [27]. These results are promising and would therefore need validation by other groups, including robust studies with different tumors as well as parallel studies on the immune response of these alternate stem cells.

## Exosomes

Exosomes are 30-100 nm specialized vesicles derived from the endocytic compartment [28]. These vesicles contain RNA, lipids and proteins and are secreted by all types of cells and can be found in body fluids such as plasma and ascites fluid [29]. There are several methods to purify exosomes. Regardless of the method, the purity can be assessed by the lack serum proteins. Apart from size, micro vesicles should not contain intracellular proteins such as mitochondrial and endosomal types. Thus, it is generally accepted that the characterization of exosomes to include analyses for endosomal proteins such as the tetraspanins. The contents of exosomes are growing and can be accessed in a public database, ExoCarta (http://www.exocarta.org).

Exosomes could mediate communication between cells through the transfer of RNA, but mostly ncRNA, in short and long distances [30]. The exosomes can enter cells through the lipid bilayer or specific receptors that are loaded in the exosomes in the originating cells. Recent studies reported on the expression of the gap junctional protein, connexin 43, on exosomes [31]. Connexin 43 can form gap junction with cells thereby allowing the vesicles to establish a paracrine transfer of molecules from the originating cell to another cell. This is possible considering miRNA has been shown to pass through connexin 43-mediated gap junction [32]. The exosomes entering the new cells could process the pre-miRNA independent of the endogenous molecular machinery of the host cells due to the presence of processing proteins within the incoming vesicles [33].

Macrophages, which are important in several diseases such as BC pathology, can communicate with the dysfunctional cells by releasing exosomes. In the case of BC, tumor-associated macrophages have been reported to transfer specific miRNA into BCCs to promote cell migration and invasion [34]. The communication between the dysfunctional cells and cells of the microenvironment is a two-way process. BCCs can initiate the communication to the surrounding cells by transferring exosomes [35, 36]. The initiating exosomes could be a method by which the BCCs establish a pre-metastatic niche to promote colonization of the tumor cells in the prepared environment [36, 37]. This two-way communication needs to be understood to develop cellular delivery of RNA to reverse the pathological process. Specifically, if the contents of the exosomes are identified, strategies could be developed to block the candidate molecule (protein or ncRNA). This would require the incorporation of several other fields to develop methods to block exosome-containing molecules. This could be the delivery of ncRNA within stem cells and nanoparticles.

## Gap Junctional Intercellular Communication (GJIC)

As discussed above, GJIC can be formed between exosomes and cells, resulting in the transfer of RNA [31]. Small ncRNA such as miRNA can also be transferred through direct intercellular communication by the formation of GJIC [38]. GJIC is formed by tow hemi-channels, each comprises of six connexins [38]. The channels vary in size depending on the type of connexin. Regardless, ions and miRNAs have been shown to pass through the gap junctions [32, 38]. The formation of GJIC between cancer cells such as BCCs and with cells of the microenvironment has become a model system to understand metastasis and dormancy [32]. GJIC is important to normal physiological functions, such as hematopoietic regulation [39]. Thus, direct targeting of the connexins is expected to be toxic to endogenous physiological functions. Thus, the model system of GJIC could lead to the molecular mechanisms by which BCCs form GJIC for the identification of targets specific to the BCCs with minimum toxicity to the endogenous cells/functions.

## Drug Delivery – Potential Approaches and Gaps

This section will discuss how MSCs could be used to interrupt the communication between BCCs and the microenvironment. The more immature BCC subset, notably, the CSCs, communicate with cells of the bone marrow microenvironment through GJIC [32, 40, 41]. This type of communication was demonstrated in a model of BC dormancy in bone marrow [32, 41]. The method is now used as a model of BC dormancy is being used in our lab to test how to use MSCs to deliver anti-miRNAs to reverse dormancy for sensitivity to chemotherapy. In another model of cancer, studies with resistance glioblastoma multiform cells tested how MSCs can serve as a cellular vehicle to deliver anit-miRNA to reverse drug resistance [42, 43]. Indeed, the studies showed that MSCs can transfer anti-miRNA through exosomes released from MSCs and, through the ability of the MSCs to form GJIC with the resistant cancer cells through GJIC [42, 43]. We have found that the MSC-derived exosomes can load the intended anti-miRNA for transfer to the target resistant glioblastoma cells.

Also, cytokines can be supportive and suppressive to tumor growth. Regardless of the role, MSCs can be used to deliver cytokines, antagonizing small molecules or shRNA to the tumors. These types of studies are feasible since the procedure has been experimentally demonstrated [44]. The engineered MSCs, also referred as second generation MSCs, can be attracted to tumors, which secrete inflammatory proteins and other factors that act as chemo attractants [45, 46].

The problem is that MSCs can also promote the tumor growth and at the same time protect the cancer cells from the innate and adaptive immune system [15, 47]. At the site of a tumor, MSCs can transfer miRNA to the tumor to regulate other functions such as modulating the production of angiogenic factors [48]. It is these homing and RNA-transfer method that the scientific community has proposed MSCs as cellular vehicle to deliver synthetic ncRNA for cancer treatment [49]. The MSCs can be used to block the tumor cells' interaction with cells of the microenvironment, such as the tumor-associated macrophages (TAM's) that can transfer miRNAs to the BCCs [34]. In this case, the MSCs will need to be `loaded' with synthetic anti-miRNAs or with a lentivirus expressing the anti-miRNAs.

BC dormancy in the bone marrow has been a subject of intense research by our group. This time of the cancer is critical because the most primitive BCCs with stem cell property survive [41]. These cells will be difficult to target with the current chemotherapies. However, the microenvironment of the bone marrow facilitates the survival of the dormant BCCs [15, 41]. It is therefore important to understand how the endogenous bone marrow cells support the dormancy phenotype so that MSCs can be used to deliver intervening molecules and ncRNA to reverse the dormancy for sensitivity to chemotherapy.

There are several sources of MSCs that can be used as off-the-shelf cells to deliver drugs for ncRNA. However, at this time, it is unclear if all MSCs have similar homing abilities. Studies similar to those that compare the immune functions of the different MSCs need to be conducted on the homing abilities, in the presence or absence of injuries [50].

There are several Phase I clinical trials using exosomes, which showed tolerance over a 21-month period. Delivering exosomes to targeted cancer cells is not a trivial matter since it remains difficult for direct targeting. Thus, the delivery of exosomes in MSCs remains a viable option.

## CONCLUSION

The unique immune properties of MSCs make them attractive therapeutic agents to deliver drugs and ncRNA to treat tumors. MSCs are attractive cells for drug delivery. They are easy to harvest and can be expanded several folds in the laboratory. Regardless of the source, MSCs seems to be similar by phenotype. However, their homing ability and their ability to form GJIC and secrete exosomes will have to be studied with robust experimental approaches.

As shown in Fig. (**1**), when the MSCs deliver drugs, they will not survive a long time and will be cleared by the innate and adaptive immune system. This is an important issue to avoid the MSCs to linger in tissues where they may form undesirable cells. If methods can be developed for effective use of MSCs in drug delivery to cancer, this could reduce the cost of healthcare because the cells will be available as off-the-shelf for immediate use in patients.

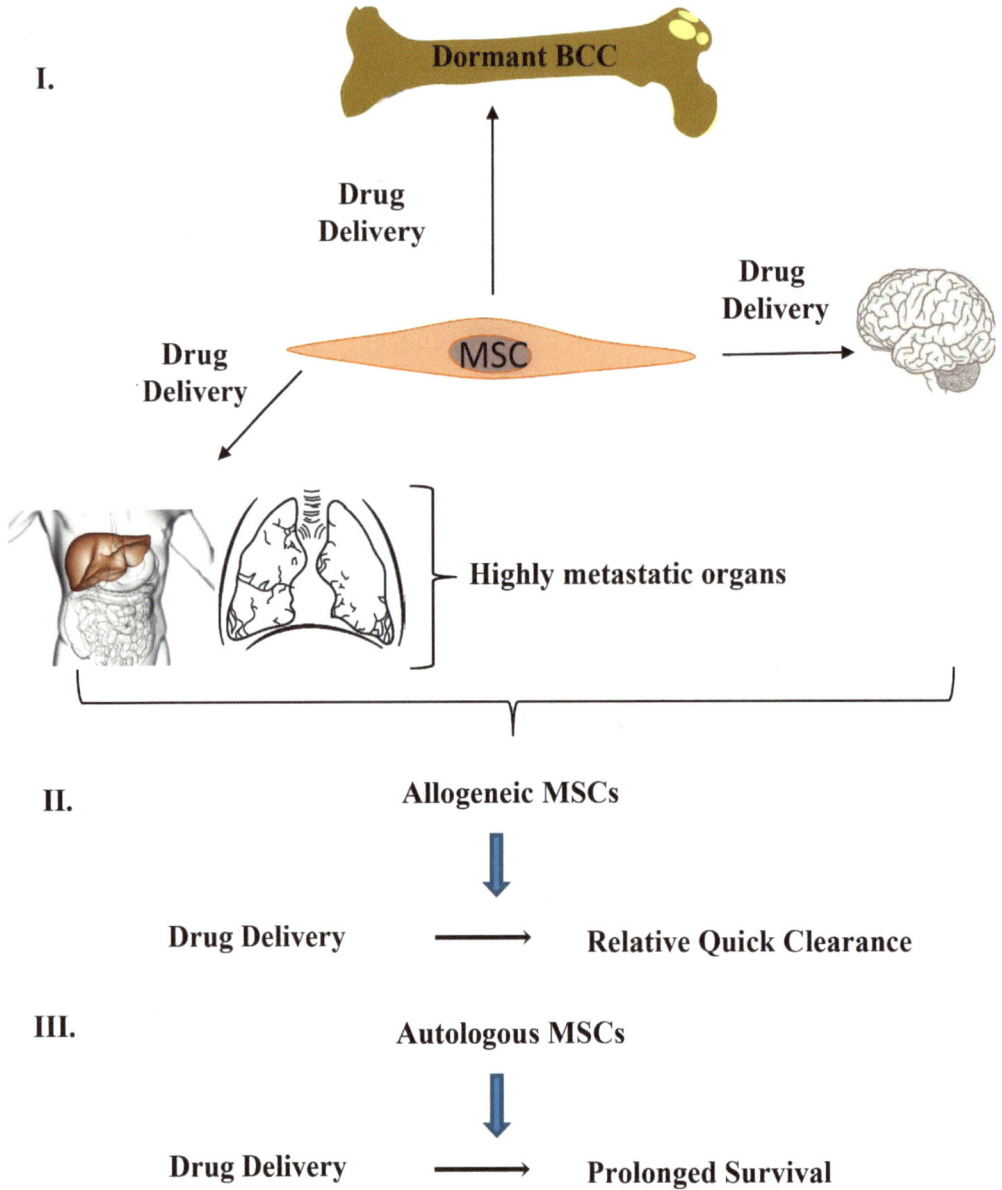

**Fig. (1).** Shown are the key issues in using MSCs as cellular vehicle of drug delivery. **I.** MSCs can be used to deliver drugs (including ncRNA) to different organs, including to the bone marrow to reverse dormancy. **II.** If allogeneic MSCs are used the cells will be cleared relatively quickly after drug delivery. **III.** If autologous MSCs are used in drug delivery the cells will remain with the drugs for prolonged period.

Overall, we return to our introduction when we discuss the problems with drug trials, the use of MSCs would be an advantage where the drugs could be delivered in a more targeted approach. This could occur within a milieu in which the MSCs are communicating with the microenvironment. The MSCs could deliver additional benefit. This latter point will need further studies to combine the desired drug for delivery, the endogenous property of MSCs and perhaps reduced toxicity with low-dose chemotherapy.

## CONFLICT OF INTEREST

The authors confirm that the authors have no conflict of interest to declare for this publication.

## ACKNOWLEDGEMENTS

Declared none.

## REFERENCES

[1]     Rameshwar P. Current thoughts on the therapeutic potential of stem cell. Methods Mol Biol 2012; 879: 3-26.
        [http://dx.doi.org/10.1007/978-1-61779-815-3_1] [PMID: 22610550]

[2]     Mariotti V, Greco SJ, Mohan RD, Nahas GR, Rameshwar P. Stem cell in alternative treatments for brain tumors: potential for gene delivery. Mol Cell Ther 2014; 2: 24.
        [http://dx.doi.org/10.1186/2052-8426-2-24] [PMID: 26056591]

[3]     Charbord P. Bone marrow mesenchymal stem cells: historical overview and concepts. Hum Gene Ther 2010; 21(9): 1045-56.
        [http://dx.doi.org/10.1089/hum.2010.115] [PMID: 20565251]

[4]     Potian JA, Aviv H, Ponzio NM, Harrison JS, Rameshwar P. Veto-like activity of mesenchymal stem cells: functional discrimination between cellular responses to alloantigens and recall antigens. J Immunol 2003; 171(7): 3426-34.
        [http://dx.doi.org/10.4049/jimmunol.171.7.3426] [PMID: 14500637]

[5]     Castillo MD, Trzaska KA, Greco SJ, Ponzio NM, Rameshwar P. Immunostimulatory effects of mesenchymal stem cell-derived neurons: implications for stem cell therapy in allogeneic transplantations. Clin Transl Sci 2008; 1(1): 27-34.
        [http://dx.doi.org/10.1111/j.1752-8062.2008.00018.x] [PMID: 20443815]

[6]     Fibbe WE, Dazzi F, LeBlanc K. MSCs: science and trials. Nat Med 2013; 19(7): 812-3.
        [http://dx.doi.org/10.1038/nm.3222] [PMID: 23836217]

[7]     Bernardo ME, Fibbe WE. Mesenchymal stromal cells and hematopoietic stem cell transplantation. Immunol Lett 2015; 168(2): 215-21.

[http://dx.doi.org/10.1016/j.imlet.2015.06.013] [PMID: 26116911]

[8]     Castro-Manrreza ME, Montesinos JJ. Immunoregulation by mesenchymal stem cells: Biological aspects and clinical applications. J Immunol Res 2015. 394917.

[9]     Chen X, Wang C, Yin J, Xu J, Wei J, Zhang Y. Efficacy of mesenchymal stem cell therapy for steroid-refractory acute graft-*versus*-host disease following allogeneic hematopoietic stem cell transplantation: a systematic review and meta-analysis. PLoS One 2015; 10(8): e0136991.
        [http://dx.doi.org/10.1371/journal.pone.0136991] [PMID: 26323092]

[10]    Lisignoli G, Cristino S, Piacentini A, Cavallo C, Caplan AI, Facchini A. Hyaluronan-based polymer scaffold modulates the expression of inflammatory and degradative factors in mesenchymal stem cells: Involvement of Cd44 and Cd54. J Cell Physiol 2006; 207(2): 364-73.
        [http://dx.doi.org/10.1002/jcp.20572] [PMID: 16331675]

[11]    Aleynik A, Gernavage KM, Mourad YSh, *et al.* Stem cell delivery of therapies for brain disorders. Clin Transl Med 2014; 3: 24.
        [http://dx.doi.org/10.1186/2001-1326-3-24] [PMID: 25097727]

[12]    Nahas G, Bliss SA, Sinha G, Ganta T, Greco SJ, Rameshwar P. Is reduction of tumor burden sufficient for the 21$^{st}$ century? Cancer Lett 2015; 356(2 Pt A): 149-55.
        [http://dx.doi.org/10.1016/j.canlet.2014.03.002] [PMID: 24632530]

[13]    Patel SA, Dave MA, Bliss SA, *et al.* T/Th17 polarization by distinct subsets of breast cancer cells is dictated by the interaction with mesenchymal stem cells. J Cancer Stem Cell Res 2014; 21003.

[14]    Helmy KY, Patel SA, Nahas GR, Rameshwar P. Cancer immunotherapy: accomplishments to date and future promise. Ther Deliv 2013; 4(10): 1307-20.
        [http://dx.doi.org/10.4155/tde.13.88] [PMID: 24116914]

[15]    Patel SA, Meyer JR, Greco SJ, Corcoran KE, Bryan M, Rameshwar P. Mesenchymal stem cells protect breast cancer cells through regulatory T cells: role of mesenchymal stem cell-derived TGF-beta. J Immunol 2010; 184(10): 5885-94.
        [http://dx.doi.org/10.4049/jimmunol.0903143] [PMID: 20382885]

[16]    Ankrum JA, Ong JF, Karp JM. Mesenchymal stem cells: immune evasive, not immune privileged. Nat Biotechnol 2014; 32(3): 252-60.
        [http://dx.doi.org/10.1038/nbt.2816] [PMID: 24561556]

[17]    Rameshwar P. Current thoughts on the therapeutic potential of stem cell. Methods Mol Biol 2012; 879: 3-26.
        [http://dx.doi.org/10.1007/978-1-61779-815-3_1] [PMID: 22610550]

[18]    Droujinine IA, Eckert MA, Zhao W. To grab the stroma by the horns: from biology to cancer therapy with mesenchymal stem cells. Oncotarget 2013; 4(5): 651-64.
        [http://dx.doi.org/10.18632/oncotarget.1040] [PMID: 23744479]

[19]    Kuraitis D, Ruel M, Suuronen EJ. Mesenchymal stem cells for cardiovascular regeneration. Cardiovasc Drugs Ther 2011; 25(4): 349-62.
        [http://dx.doi.org/10.1007/s10557-011-6311-y] [PMID: 21637968]

[20]    Torrente Y, Polli E. Mesenchymal stem cell transplantation for neurodegenerative diseases. Cell Transplant 2008; 17(10-11): 1103-13.

[http://dx.doi.org/10.3727/096368908787236576] [PMID: 19181205]

[21]    Amara I, Touati W, Beaune P, de Waziers I. Mesenchymal stem cells as cellular vehicles for prodrug gene therapy against tumors. Biochimie 2014; 105: 4-11.
[http://dx.doi.org/10.1016/j.biochi.2014.06.016] [PMID: 24977933]

[22]    Collet G, Grillon C, Nadim M, Kieda C. Trojan horse at cellular level for tumor gene therapies. Gene 2013; 525(2): 208-16.
[http://dx.doi.org/10.1016/j.gene.2013.03.057] [PMID: 23542073]

[23]    Tanna T, Sachan V. Mesenchymal stem cells: potential in treatment of neurodegenerative diseases. Curr Stem Cell Res Ther 2014; 9(6): 513-21.
[http://dx.doi.org/10.2174/1574888X09666140923101110] [PMID: 25248677]

[24]    Allers C, Jones JA, Lasala GP, Minguell JJ. Mesenchymal stem cell therapy for the treatment of amyotrophic lateral sclerosis: signals for hope? Regen Med 2014; 9(5): 637-47.
[http://dx.doi.org/10.2217/rme.14.30] [PMID: 25372079]

[25]    Ahmed AU, Lesniak MS. Glioblastoma multiforme: can neural stem cells deliver the therapeutic payload and fulfill the clinical promise? Expert Rev Neurother 2011; 11(6): 775-7.
[http://dx.doi.org/10.1586/ern.11.65] [PMID: 21651324]

[26]    Semi K, Matsuda Y, Ohnishi K, Yamada Y. Cellular reprogramming and cancer development. Int J Cancer 2013; 132(6): 1240-8.
[http://dx.doi.org/10.1002/ijc.27963] [PMID: 23180619]

[27]    Zhao Q, Gregory CA, Lee RH, *et al.* MSCs derived from iPSCs with a modified protocol are tumor-tropic but have much less potential to promote tumors than bone marrow MSCs. Proc Natl Acad Sci USA 2015; 112(2): 530-5.
[http://dx.doi.org/10.1073/pnas.1423008112] [PMID: 25548183]

[28]    Hurley JH. ESCRTs are everywhere. EMBO J 2015; 34(19): 2398-407.
[http://dx.doi.org/10.15252/embj.201592484] [PMID: 26311197]

[29]    Raposo G, Stoorvogel W. Extracellular vesicles: exosomes, microvesicles, and friends. J Cell Biol 2013; 200(4): 373-83.
[http://dx.doi.org/10.1083/jcb.201211138] [PMID: 23420871]

[30]    Valadi H, Ekström K, Bossios A, Sjöstrand M, Lee JJ, Lötvall JO. Exosome-mediated transfer of mRNAs and microRNAs is a novel mechanism of genetic exchange between cells. Nat Cell Biol 2007; 9(6): 654-9.
[http://dx.doi.org/10.1038/ncb1596] [PMID: 17486113]

[31]    Soares AR, Martins-Marques T, Ribeiro-Rodrigues T, *et al.* Gap junctional protein Cx43 is involved in the communication between extracellular vesicles and mammalian cells. Sci Rep 2015; 5: 13243.
[http://dx.doi.org/10.1038/srep13243] [PMID: 26285688]

[32]    Lim PK, Bliss SA, Patel SA, *et al.* Gap junction-mediated import of microRNA from bone marrow stromal cells can elicit cell cycle quiescence in breast cancer cells. Cancer Res 2011; 71(5): 1550-60.
[http://dx.doi.org/10.1158/0008-5472.CAN-10-2372] [PMID: 21343399]

[33]    Melo SA, Sugimoto H, OConnell JT, *et al.* Cancer exosomes perform cell-independent microRNA biogenesis and promote tumorigenesis. Cancer Cell 2014; 26(5): 707-21.

[http://dx.doi.org/10.1016/j.ccell.2014.09.005] [PMID: 25446899]

[34]   Yang M, Chen J, Su F, *et al.* Microvesicles secreted by macrophages shuttle invasion-potentiating microRNAs into breast cancer cells. Mol Cancer 2011; 10: 117.
[http://dx.doi.org/10.1186/1476-4598-10-117] [PMID: 21939504]

[35]   Dutta S, Warshall C, Bandyopadhyay C, Dutta D, Chandran B. Interactions between exosomes from breast cancer cells and primary mammary epithelial cells leads to generation of reactive oxygen species which induce DNA damage response, stabilization of p53 and autophagy in epithelial cells. PLoS One 2014; 9(5): e97580.
[http://dx.doi.org/10.1371/journal.pone.0097580] [PMID: 24831807]

[36]   Fong MY, Zhou W, Liu L, *et al.* Breast-cancer-secreted miR-122 reprograms glucose metabolism in premetastatic niche to promote metastasis. Nat Cell Biol 2015; 17(2): 183-94.
[http://dx.doi.org/10.1038/ncb3094] [PMID: 25621950]

[37]   Kosaka N, Iguchi H, Hagiwara K, Yoshioka Y, Takeshita F, Ochiya T. Neutral sphingomyelinase 2 (nSMase2)-dependent exosomal transfer of angiogenic microRNAs regulate cancer cell metastasis. J Biol Chem 2013; 288(15): 10849-59.
[http://dx.doi.org/10.1074/jbc.M112.446831] [PMID: 23439645]

[38]   Su V, Lau AF. Connexins: mechanisms regulating protein levels and intercellular communication. FEBS Lett 2014; 588(8): 1212-20.
[http://dx.doi.org/10.1016/j.febslet.2014.01.013] [PMID: 24457202]

[39]   Foss B, Hervig T, Bruserud O. Connexins are active participants of hematopoietic stem cell regulation. Stem Cells Dev 2009; 18(6): 807-12.
[http://dx.doi.org/10.1089/scd.2009.0086] [PMID: 19355839]

[40]   Bliss SA, Greco SJ, Rameshwar P. Hierarchy of breast cancer cells: key to reverse dormancy for therapeutic intervention. Stem Cells Transl Med 2014; 3(7): 782-6.
[http://dx.doi.org/10.5966/sctm.2014-0013] [PMID: 24833590]

[41]   Patel SA, Ramkissoon SH, Bryan M, *et al.* Delineation of breast cancer cell hierarchy identifies the subset responsible for dormancy. Sci Rep 2012; 2: 906.
[http://dx.doi.org/10.1038/srep00906] [PMID: 23205268]

[42]   Munoz JL, Bliss SA, Greco SJ, Ramkissoon SH, Ligon KL, Rameshwar P. Delivery of functional anti-miR-9 by mesenchymal stem cell-derived exosomes to glioblastoma multiforme cells conferred chemosensitivity. Mol Ther Nucleic Acids 2013; 2: e126.
[http://dx.doi.org/10.1038/mtna.2013.60] [PMID: 24084846]

[43]   Munoz JL, Rodriguez-Cruz V, Greco SJ, Ramkissoon SH, Ligon KL, Rameshwar P. Temozolomide resistance in glioblastoma cells occurs partly through epidermal growth factor receptor-mediated induction of connexin 43. Cell Death Dis 2014; 5: e1145.
[http://dx.doi.org/10.1038/cddis.2014.111] [PMID: 24675463]

[44]   Park JH, Ryu CH, Kim MJ, Jeun SS. Combination therapy for gliomas using temozolomide and interferon-beta secreting human bone marrow derived mesenchymal stem cells. J Korean Neurosurg Soc 2015; 57(5): 323-8.
[http://dx.doi.org/10.3340/jkns.2015.57.5.323] [PMID: 26113958]

[45]   Dwyer RM, Potter-Beirne SM, Harrington KA, *et al.* Monocyte chemotactic protein-1 secreted by primary breast tumors stimulates migration of mesenchymal stem cells. Clin Cancer Res 2007; 13(17): 5020-7.
[http://dx.doi.org/10.1158/1078-0432.CCR-07-0731] [PMID: 17785552]

[46]   Spaeth E, Klopp A, Dembinski J, Andreeff M, Marini F. Inflammation and tumor microenvironments: defining the migratory itinerary of mesenchymal stem cells. Gene Ther 2008; 15(10): 730-8.
[http://dx.doi.org/10.1038/gt.2008.39] [PMID: 18401438]

[47]   Reagan MR, Kaplan DL. Concise review: Mesenchymal stem cell tumor-homing: detection methods in disease model systems. Stem Cells 2011; 29(6): 920-7.
[http://dx.doi.org/10.1002/stem.645] [PMID: 21557390]

[48]   Lee JK, Park SR, Jung BK, *et al.* Exosomes derived from mesenchymal stem cells suppress angiogenesis by down-regulating VEGF expression in breast cancer cells. PLoS One 2013; 8(12): e84256.
[http://dx.doi.org/10.1371/journal.pone.0084256] [PMID: 24391924]

[49]   Lee HK, Finniss S, Cazacu S, *et al.* Mesenchymal stem cells deliver synthetic microRNA mimics to glioma cells and glioma stem cells and inhibit their cell migration and self-renewal. Oncotarget 2013; 4(2): 346-61.
[http://dx.doi.org/10.18632/oncotarget.868] [PMID: 23548312]

[50]   Wang Q, Yang Q, Wang Z, *et al.* Comparative analysis of human mesenchymal stem cells from fetal-bone marrow, adipose tissue, and Wartons jelly as sources of cell immunomodulatory therapy. Hum Vaccin Immunother 2016; 12(1): 85-96.
[http://dx.doi.org/10.1080/21645515.2015.1030549] [PMID: 26186552]

# SUBJECT INDEX

www.ingramcontent.com/pod-product-compliance
Lightning Source LLC
Chambersburg PA
CBHW041707210326
41598CB00007B/564